Degradation and Recycling of Polymer Materials

Degradation and Recycling of Polymer Materials

Guest Editor

Hisayuki Nakatani

MDPI

Basel • Beijing • Wuhan • Barcelona • Belgrade • Novi Sad • Cluj • Manchester

Guest Editor
Hisayuki Nakatani
Polymeri Materials
Laboratory
Nagasaki University
Nagasaki
Japan

Editorial Office
MDPI AG
Grossspeteranlage 5
4052 Basel, Switzerland

This is a reprint of the Special Issue, published open access by the journal *Polymers* (ISSN 2073-4360), freely accessible at: https://www.mdpi.com/journal/polymers/special_issues/8O3NL2DN9J.

For citation purposes, cite each article independently as indicated on the article page online and as indicated below:

Lastname, A.A.; Lastname, B.B. Article Title. *Journal Name* **Year**, *Volume Number*, Page Range.

ISBN 978-3-7258-4701-3 (Hbk)
ISBN 978-3-7258-4702-0 (PDF)
https://doi.org/10.3390/books978-3-7258-4702-0

polymers

MDPI

Article

Differences in the Residual Behavior of a Bumetrizole-Type Ultraviolet Light Absorber during the Degradation of Various Polymers

Hisayuki Nakatani [1,2,*], Taishi Uchiyama [1], Suguru Motokucho [1,2], Anh Thi Ngoc Dao [1], Hee-Jin Kim [3], Mitsuharu Yagi [3] and Yusaku Kyozuka [2]

1 Polymeri Materials Laboratory, Chemistry and Materials Engineering Program, Nagasaki University, 1-14 Bunkyo-machi, Nagasaki 852-8521, Japan; bb52122606@ms.nagasaki-u.ac.jp (T.U.); motoku@nagasaki-u.ac.jp (S.M.); anh.dao@nagasaki-u.ac.jp (A.T.N.D.)

2 Organization for Marine Science and Technology, Nagasaki University, 1-14 Bunkyo-machi, Nagasaki 852-8521, Japan; kyozuka@nagasaki-u.ac.jp

3 Graduate School of Fisheries and Environmental Sciences, Nagasaki University, 1-14 Bunkyo-machi, Nagasaki 852-8521, Japan; heejin@nagasaki-u.ac.jp (H.-J.K.); yagi-m@nagasaki-u.ac.jp (M.Y.)

* Correspondence: h-nakatani@nagasaki-u.ac.jp

Abstract: The alteration of an ultraviolet light absorber (UVA: UV-326) in polymers (PP, HDPE, LDPE, PLA, and PS) over time during degradation was studied using an enhanced degradation method (EDM) involving sulfate ion radicals in seawater. The EDM was employed to homogeneously degrade the entire polymer samples containing the UVA. The PP and PS samples containing 5-phr (phr: per hundred resin) UVA films underwent rapid whitening, characterized by the formation of numerous grooves or crushed particles. Notably, the UVA loss rate in PS, with the higher glass transition temperature (Tg), was considerably slower. The behavior of crystalline polymers, with the exception of PS, was analogous in terms of the change in UVA loss rate over the course of degradation. The significant increase in the initial loss rate observed during EDM degradation was due to microplasticization. A similar increase in microplasticization rate occurred with PS; however, the intermolecular interaction between UVA and PS did not result in as pronounced an increase in loss rate as observed in other polymers. Importantly, the chemical structure of UVA remained unaltered during EDM degradation. These findings revealed that the primary cause of UVA loss was leaching from the polymer matrix.

Keywords: microplastics; ultraviolet light absorber; bleaching; polypropylene; polystyrene; polyethylene

1. Introduction

Plastic debris has been discharged into the ocean, resulting in the widespread distribution of microplastic (MP) pollution [1–8]. Reflecting the substantial production volume of commercial plastics, MPs are primarily comprised of polyethylene (PE), polypropylene (PP), and polystyrene (PS) particles that float on the sea surface. Given their non-biodegradable nature, these polymeric products persist in the marine environment for extended periods. Initial investigations proposed the transfer of chemicals from plastic debris ingested by sea birds to their tissues [9]. The evaluation of chemical transfer from marine plastic debris such as MPs to other organisms in the oligotrophic open ocean is yet to be elucidated. However, due to the size of micrometers, MPs are deemed bioavailable across the entire food web, calling for a consideration of chemical transfer at the bottom of the food chain where bioaccumulation could potentially initiate.

MPs can form in diverse locations, with some generated in the terrestrial region (on land) and others in the sea. Occasionally, MPs partially transition from the sea to the atmosphere, circulating in various places. Halle et al. outlined the mechanism of MP formation in the sea, linking it to sunlight exposure [4]. A total of 92% of the collected samples in the study were made of PE, the type of polymer that is principally degraded by photo-oxidation, in which UV radiation triggers free radical reactions in the polymeric chain [10]. In studying photo-oxidation degradation, the broadening of the carbonyl groups (measured by Fourier transform infrared spectroscopy, or FT-IR), particularly the ketone, carboxylic acid, and ester functional groups formed during UV irradiation, is commonly used to assess the progression of weathering within the 1650–1850 cm^{-1} interval. The carbonyl index (or CI) obtained from such measurements works as a metric for characterizing the advancement of polymer oxidation. The MPs in Halle's study exhibited carbonyl indexes ranging from 0.13 to 0.74, signifying advanced stages of oxidation. While the detailed mechanism for fragmentation remained unclear, it was evident that sunlight, especially in the visible and/or UV regions, played a role in MP formation.

The substantial presence of large quantities of MPs raises concerns about potential risks to marine ecosystems. However, the quantitative impact of MPs on marine life remains uncertain, and there is no definite answer to the numerous concerns surrounding this issue. Plastic products, which are rarely used alone and often contain various types of organic or inorganic compounds, and among them, organic additives [11–14], are of particular concerns. These additives may undergo transformation during MP formation. Recently, Z. Tian et al. reported the induction of acute mortality in coho salmon by a highly toxic quinone transformation product of N-(1,3-dimethylbutyl)-N′-phenyl-p-phenylenediamine (6PPD), a globally ubiquitous tire rubber antioxidant [15]. The toxicity of 6PPD-quinone to juvenile coho salmon was significant, with a median lethal concentration (LC_{50}) of 0.79 ± 0.16 μg/liter. Throughout the product life cycle, antioxidants such as 6PPD are specifically engineered to diffuse to tire rubber surfaces. Their primary function is to scavenge atmospheric and ground-level ozone, creating protective films that eliminate the ozone-mediated oxidation of rubber elastomers [16]. Consequently, the intentional reaction of 6PPD in tire rubbers leads to the formation of 6PPD-quinone and related transformation products, inadvertently elevating toxicity and environmental risk.

In the study of the ecological impact of MPs, critical questions arise: How much of the additives leach out during MP formation? What changes occur in the chemical structures of additives? Additionally, how does the additive behavior of different plastics vary during MP formation? The above-mentioned case of 6PPD highlights the transformation of a seemingly safe additive into a toxic compound in nature, emphasizing the importance of understanding chemical changes during MP formation for assessing its impact on living organisms. Notably, UV stabilizers (UVAs), essential additives in plastic substances, particularly for outdoor packaging and corrosion inhibition, warrant detailed investigation after the degradation of containing plastics. Bumetrizole-type UVA, frequently used in commodity plastics such as PP and PE, shares a chemical structure similarity with 6PPD, and its degraded product forms a quinone body akin to 6PPD. Thoroughly investigating UVA changes post degradation is essential due to potential ecological impacts.

As mentioned above, the primary locations of MP formation are considerably complicated, occurring both on land and in the sea, and each is governed by distinct mechanisms [17]. Surface observations of MPs retrieved from the seashore reveal delamination, whereas those from the riverside exhibit an abrasion patch structure [17]. The delaminated part, releasing smaller MPs into the sea, is a crucial aspect. Although MP formation occurs both on land and in the sea, it is considered to primarily be produced in the sea and its vicinity. Consequently, reference samples formed in the sea are deemed more suitable for investigating the leaching and alteration behavior of UVA during MP formation. Delamination, achieved by exposure to visible and/or UV light in water, has been utilized by several researchers to obtain MP reference samples [18–20]. Similarly, in our previous study, we conducted a photodegradation test of PP film in water under visible light irradiation using a specific photocatalyst, resulting in planar delamination and acquisition of the MP reference sample [21]. While MP formation is inherently linked to autoxidation, the rates induced by light irradiation, even with a photocatalyst, were notably sluggish. To facilitate a comparative study of additive behavior on different plastics during MP formation, the development of an accelerated MP production method is imperative [22,23]. In our previous study [24], we degraded PP in seawater using a sulfate ion radical ($SO_4\bullet^-$) as a highly efficient initiator for plastic degradation. This combination, coupled with pH control, significantly accelerated the degradation process. Notably, this approach was effective for the accelerated degradation of other C-C bonded polymers such as HDPE, LDPE, PLA, and PS, which share the same degradation mechanism (autoxidation).

In this study, we studied the quality-altering behavior of bumetrizole-type UVA (UV-326) in PP, HDPE, LDPE, PLA, and PS over time during degradation using an enhanced degradation method (EDM) employing a sulfate ion radical in seawater. UVA concentration in these polymers was quantified through pyrolysis gas chromatography/mass spectrometry (Py-GC/MS), and structural changes in UVA were confirmed through FT-IR measurements.

2. Materials and Methods

2.1. Materials

PP was supplied by Prime Polymer Co., Ltd., Tokyo, Japan (product name: J-700GP). The melt flow rate (MFR) and density were 8 g/10 min and 0.9 g/cm^3. PLA was supplied by Mitsui Chemicals, Inc., Tokyo, Japan (production name: Gread H-100). The weight-average molecular weight (Mw) and molecular weight distribution (Mw/Mn) were 1.2×10^5 and 1.1, respectively. PS was purchased from Sigma-Aldrich Co. LLC(St. Louis, MO, USA) The weight-average molecular weight (Mw) and molecular weight distribution (Mw/Mn) were 3.5×10^5 and 2.1, respectively. HDPE and LDPE were purchased from Sigma-Aldrich Co. LLC. The melt index (190 °C/2.16 kg) values of HDPE and LDPE were 10 g/10 min and 25 g/10 min, respectively. Bumetrizole (UVA: UV-326) and potassium persulfate ($K_2S_2O_8$) were purchased from Wako Pure Chemical Industries (Osaka, Japan) Seawater was prepared with Gex artificial saltwater purchased from Amazon.co.jp.

2.2. Preparation of Polymer Samples Containing Ultraviolet Light Absorber (UVA)

The polymer and UVA blend were prepared using an Imoto Seisakusyo IMC-1884 melting mixer (Kyoto, Japan). The polymer pellets (ca. ϕ5 mm) and bumetrizole fine particles (ca. 0.01 mm × 0.1 mm) were employed as polymer and UVA, respectively. Next, 2 g-powdery of polymer sample and 0.1 g UVA were put into a 50 mL glass vessel and then premixed. The mixture was put into the melting mixer, and melt mixing was performed at 180 °C and 50 rpm for 6 min. The obtained sample was molded into a film (ca. 50 × 50 × 0.075 mm) by compression molding at 180 °C under 10 MPa for 5 min. The film was cut to 5 mm × 5 mm × 0.075 mm in size and was employed as polymer film containing 5-phr (phr: per hundred resin) UVA. For PS samples, 10-phr UVA-containing films were prepared in a similar method.

2.3. Degradation Using Sulfate Ion Radicals in Pure Water

The polymer film samples, in their as-prepared state, were employed for the degradation procedure, following the methodology outlined in our previously published report [22]. In this process, five pieces of each film sample were put into individual 100 mL glass vessels, each containing a 20 mL aqueous solution with 0.54 g $K_2S_2O_8$. The vessels were maintained at ca. 65 °C for a duration of 12 h, with continuous stirring at a tip speed of ca. 100 rpm. To account for the consumption of the oxidant, an equivalent amount of $K_2S_2O_8$ aqueous solution was renewed every 12 h. This treatment was repeated for various time periods, subjecting the films to a continuous degradation process. The degradation was performed in pure water using sulfate ion radicals.

2.4. Degradation Using Sulfate Ion Radicals in Seawater (Enhanced Degradation Method)

The degradation was performed in seawater using a sulfate ion radical. The procedure was according to our previous report [21]. (1) Five pieces of each film were put into a

100 mL glass vessel containing a 20 mL of seawater solution with 0.54 g $K_2S_2O_8$ at ca. 65 °C for 12 h under stirring with a stirrer tip speed of ca. 100 rpm. (2) An equal amount of $K_2S_2O_8$ seawater solution was added to compensate for the consumption of oxidant, and its degradation was carried out for 12 h under the same conditions. (3) The five pieces of the film were then transferred to a new 100 mL glass vessel containing 20 mL of seawater solution with 0.54 g $K_2S_2O_8$, and the degradations were started again under the same conditions. The enhanced degradation method was carried out for a predetermined number of 15 days using (1) to (3) as one set (total 15 sets). The pH value of the solution was changed from 8.2 to 3 during each set.

2.5. Fourier Transform Infrared (FT-IR) Analysis

For the IR spectra, 16 scans were measured with an FT-IR spectrometer (Jasco FT-IR 660 plus, Tokyo, Japan) at a resolution of 4 cm^{-1} over the full mid-IR range (400–4000 cm^{-1}). The carbonyl index (CI) of PP was calculated as the band intensity ratio of the carbonyl group (ca. 1715 cm^{-1})/scissoring CH_2 group (ca. 1450 cm^{-1}). The presence of hydroperoxide was confirmed by the development of a broad peak around 3400 cm^{-1}.

2.6. Pyrolysis Gas Chromatography/Mass (Py-GC/MS) Spectroscopy Measurement and Creation of Calibration Curve

A multi-functional pyrolyzer (Frontier Labs., Fukusima, Japan, EGA/PY-3030D) was attached to a GC/MS spectroscope (SHIMADZU, Kyoto, Japan, GCMS-QP2010 PLUS). The measurement was performed on a 100 μg sample. The pyrolysis was performed at 550 °C. Helium was used as the carrier gas for the capillary column with a flow rate of 1.0 mL/min. The MS system was operated under electron ionization mode at 70 eV.

The UVA (UV-326) concentration in various polymers was measured by the Py-GC/MS. The UVA (UV-326) peaks were identified and assigned in the Py-GC/MS by comparing them to the UVA-free polymer samples. As shown in Figure S1, the UVA peak appeared at a retention time of 23.5 to 24.5 min. The area ratio calculated by dividing the peak area by the total area was plotted against known UVA concentrations to create a calibration curve (see Figure S2). The UVA concentrations in various polymers at the time of degradation were calculated using this calibration curve based on area ratios obtained from the py-GC/MS measurements.

2.7. Scanning Electron Microscope (SEM) with Energy Dispersive X-Ray Spectroscopy Analysis

The SEM/EDX analysis was carried out with a JSM-7500FAM (JEOL, Tokyo, Japan) at 5.0 kV. The working distance was about 3 × 4 mm. Samples were placed in a drying oven maintained at 27 °C for 30 min and were sputter-coated with gold before SEM imaging.

2.8. Gel Permeation Chromatography (GPC) Analysis

PP samples were dissolved in 5ml of o-dichlorobenzene, and their molecular weights were directly measured using an HLC-8321GPC/HT GPC system (Tosoh Co., Ltd., Tokyo, Japan) at 140 °C.

PS samples were dissolved in 5 mL of chloroform in a small vial, and the obtained sample solution was directly measured by GPC. The molecular weight was determined by GPC (SHIMADZU, Kyoto, Japan, Prominence GPC system) at 40 °C using chloroform as a solvent.

3. Results and Discussion

3.1. Homogeneous Degradation Behavior

Weathering degradation primarily affects the polymer matrix's surface, making it a prolonged process for degradation to reach the interior. Analyzing the quality change of an encapsulated UVA additive within the polymer over a short timeframe is challenging. To facilitate the detection of UVA quality changes during degradation, a method that allows for simultaneous initiation on the surface and within the polymer matrix is essential. We employed a novel "enhanced degradation method (EDM)" using sulfate ion radicals in

polymers

MDPI

Article

Tenebrio molitor Could Be an Efficient Pre-Treatment Bioagent for Polystyrene Initial Deterioration and Further Application of *Pleurotus eryngii* and *Trametes versicolor* in Microplastic Biodegradation

Slobodan Stefanović [1], Milena Dimitrijević [2], Dragosav Mutavdžić [2], Kristina Atlagić [3], Slobodan Krnjajić [2] and Žaklina Marjanović [2,*]

[1] Faculty of Applied Ecology FUTURA, Metropolitan University, Požeška 83a, 11000 Belgrade, Serbia; slobodan.stefanovic@futura.edu.rs
[2] University of Belgrade—Institute for Multidisciplinary Research, Kneza Višeslava 1, 11000 Belgrade, Serbia; milena.dimitrijevic@imsi.bg.ac.rs (M.D.); gane@imsi.bg.ac.rs (D.M.); slobodan.krnjajic@imsi.bg.ac.rs (S.K.)
[3] Department of Physiology and Biophysics, Faculty of Biology, University of Belgrade, Studentski Trg 16, 11158 Belgrade, Serbia; kristina.tesanovic@bio.bg.ac.rs
* Correspondence: zaklina@yahoo.com or zaklina@imsi.bg.ac.rs

Abstract

Plastic is a major organic pollutant globally but has only recently been recognized for its recalcitrant nature and resistance to degradation. Although vast amounts of plastic debris are overwhelming the planet, the search for solutions to its degradation has only recently begun. One of the most well-known agents of plastic biodegradation is the larvae of *Tenebrio molitor*, which can alter the structure of polymers like polystyrene. However, while this insect can cause deterioration, its frass, which still consists of polystyrene microplastics, remains a problem. We investigated whether this frass could be further degraded by strains of white rot fungi, specifically *Pleurotus eryngii* and *Trametes versicolor*. We introduced two PS derivatives (styrofoam and stirodure) to the fungi in liquid media and evaluated oxidative metabolism enzymes (laccase, Mn-peroxidase, lignin-peroxidase) activities, and the phenolic products of the potential aromatic polymer degradation in the media. Finally, we evaluated FTIR spectra to determine if we could detect changes in polystyrene molecule degradation. Both fungi produced high amounts of enzymes, particularly when the polystyrene was present. Large quantities of phenolic substances were simultaneously detected, some associated with polystyrene degradation. FTIR spectra of different polystyrene products confirmed species-specific mechanisms for their degradation by experimental fungal strains.

Keywords: polystyrene biodegradation; microplastic biodegradation; *Tenebrio molitor*; *Pleurotus eryngii*; Trametes versicolor

1. Introduction

Plastic polymers are a major problem worldwide because they accumulate as hardly degradable solid waste on the Earth's surface [1]. There are various forms of plastics, of which polystyrene (PS) has been one of the most produced and used in the last decade. In 2016 alone, the global production of PS was around 14.7 million metric tons per year [2], while in Europe, it represents 6–7% of total annual plastic production [3]. PS production

has consistently increased due to its excellent properties for use as a building material, packaging foam, food packaging, and disposable cups, plates, and cutlery. However, most of these products, usually after a single use, sooner or later end up in landfills and nature. In its basis, PS is a large and stable polymer consisting of aromatic styrene monomers with a large molecular weight and high hydrophobicity [4,5].

PS products are easily physically disintegrated into small particles, which is why they are the most prominent producers of microplastics (MPs) in the environment. MPs usually end up in the soil around construction fields, garbage disposals, or anywhere people dump them. Like other plastics, due to the very stable C-C bonds, the degradation of PS is generally a very slow process lasting over long periods [4].

However, since the PS is an organic polymer similar to lignin, enzymatic systems of organisms like bacteria and fungi, as well as some insect larvae (e.g., *Tenebrio molitor* L.), are able to perform its biodegradation [1,6]. Biodegradation is a complex process that transforms plastic objects and particles, as well as large polymers, into smaller units mediated by biological factors [7,8]. This process is assumed to be divided into four stages: biofilm formation, biodeterioration, biofragmentation, and assimilation [9]. The first step in degrading particles made of high molecular weight and long-chain polymers is the weakening of their structure. Several factors can influence plastic biodegradation, including environmental conditions, exposure to UV light, the hydrophobicity of the exposed area, the chemical structure, crystallinity grade, and elasticity [10–12]. Organisms like microorganisms or insect larvae may initiate the biodegradation of some plastics, which, at their final stage, become a food source [6]. Thus, biodegradation is an environmentally friendly and cost-effective method of degrading plastics, in contrast to photodegradation and thermal oxidation, which are classified as environmentally harmful methods.

White rot fungi are known as the most efficient degraders of organic matter in nature, producing different enzymes that are used in diverse processes leading towards efficient and complete degradation of recalcitrant natural polymers like lignin [13–16]. Lignin is comprised of three types of phenyl–propane (aromatic) units bonded to one another in different ways [17]. In the presence of a potential substrate, white rot fungi excrete exoenzymes that attack the surface of the material and gradually degrade adequate polymers [18]. The most well-known enzymes for the biodegradation of lignin are laccases (Lac) and peroxidases, presumably Mn- (MnP), lignin- (LiP), and versatile peroxidases [19–21], which have different substrate ranges depending on their respective redox potentials [22,23]. All three enzymes are non-specific towards substrates with high redox potentials—they attack the phenolic, but also non-phenolic subunits of lignin [21,24–26]. Laccases produced by white rot basidiomycetes have high redox potential and diverse catalytic properties, so they can perform diverse biotechnologically challenging processes [27–29]. Laccases use oxygen as an electron acceptor to oxidize phenolic and non-phenolic compounds [30–32]. MnP catalyses oxidation of Mn^{2+} to Mn^{3+} in a H_2O_2-dependent reaction, attacking both phenolic and non-phenolic compounds [33], while LiP catalyses H_2O_2-dependent oxidative depolymerization of lignin [34].

Since the first step in plastic biodegradation is the formation of biofilms on the material surface, filamentous fungi that have fast hyphal apical and multi-directional growth [35] appear to be very suitable for performing such a process. Additionally, they are capable of synthesizing small globular proteins (<20 kDa) known as hydrophobins that form amphipathic monolayers at interfaces with the substrates, thus forming compact fungus–enzyme–substrate relationships [36].

Similarly to lignin degradation, in plastic biodegradation processes, the fungal mycelia overgrow the plastic and produce enzymes that break down the high molecular weight polymers into low molecular weight units, or even mineralize them totally [1]. Under the scanning electron microscope (SEM), the damage caused by fungi on the plastic surfaces can be easily detected [37,38]. White rot fungi such as *Pleurotus eryngii* (DC.) Quél. and *Trametes versicolor* (L.) Lloyd are known to be efficient not only in the biodegradation of lignin but also in the biodegradation of plastic polymers [1]. *P. eryngii* was able to degrade di-2-ethylhexyl phthalate (DEHP) plastics due to the production of MnP [39]. Similarly, *T. versicolor* degraded high molecular weight polyethylene with the enzyme MnP [40] and the lignin–polystyrene copolymer by producing oxidative enzymes [41]. Even though high resistance of the C–C backbones in PS polymer makes its degradation difficult for fungi, Milstein et al. [41] recorded SEM-confirmed polymer overgrowth, as well as biodegradation of the lignin–polystyrene copolymer by basidiomycete white rot fungi *Phanerochaete chrysosporium*, *Trametes versicolor*, and *Pleurotus ostreatus*, accompanied by the production of LiP and Lac.

However, for initiating the PS degradation process, fungi are likely not efficient because their exoenzymes are too large to penetrate deeply into the material, so they can only act on the polymer's surfaces, eventually causing surface erosion [42]. For this reason, few studies have used pre-treatments like UV light, ozonation, chemical oxidants, and thermal exposures to oxidize the polymer and create reactive functional groups (carbonyl/carboxyl/hydroxyl groups) to enable fungal enzymes to act [43,44]. Oxidative enzymes secreted by fungi gradually corrode the hydrophobic inert surface of PS, making it rough and more accessible for further colonization by fungal mycelia [45]. Additionally, in our previous work, we concluded that the efficiency of lignolytic enzymes depends on the substrate's surface—the larger the surface area of the substrate (smaller particles), the more exposed it will be to the exoenzymes, and the more efficient the degradation will be [16].

In addition to the fungal enzymes' ability to degrade plastics, the insect larvae of *T. molitor* are also capable of degrading PS in their digestive tract due to their gut microflora [6,46,47]. Styrene, the building block of PS, could be metabolized and utilized by bacteria [48], including insect gut bacteria. Insects can survive long periods (even months) solely feeding on plastics and even transform from larvae to imago stadium in such conditions (S. Krnjajić, unpublished). The darkling beetle *Plesiophthalmus davidis* larvae could survive on PS as a feed and degrade PS in their gut via depolymerization and oxidation [49]. However, after the polystyrene consumption, insects reveal frass that consists of small particles of undegraded plastic material. These particles are app. 0.2 mm in size, therefore belonging to the MP (Krnjajić et al., unpublished), implying that insects could be good bioagents for the deterioration of PS but are incapable of complete biodegradation.

The plastic biodegradation process is very slow and not easy to observe under the experimental conditions. Several studies have assessed the plastic polymer degradation abilities of various microorganisms using Fourier transform infrared spectroscopy (FTIR) [37,50–52] by comparing the spectra obtained from the plastics before and after treatment with fungi. The graphical representations of the initial spectrum and the difference spectrum show the regions/absorbances that were subjected to chemical changes, including the formation or disappearance of functional groups on the surface of a polymer upon microbial degradation [53]. The appearance of new peaks, along with the disappearance and shifting of peaks attributed to the stretching and bending vibrations of different functional groups, is assumed to confirm the structural changes that occurred on the polymer

surface upon biodegradation [37,50–52]. Additionally, it is assumed that the oxidoreductases (laccases and peroxidases) are involved in PS degradation into smaller molecules, such as oligomers, dimers, and monomers [54,55]. Similarly to lignin, these products can be detected using high-pressure liquid chromatography (HPLC) for detecting aromatic products of fungi-mediated degradation of PS [16]. This technique can be combined with FTIR spectra to demonstrate the biodegradation of PS by biological agents.

In the present study, we grew the insect larvae of *T. molitor* on two types of PS: styrofoam, expanded polystyrene (EPS), and styrodure, extruded polystyrene (XPS). The processed PS in the form of obtained larval frass, which was the size of the MP and supposedly having undergone a certain level of polymer destruction in the insect gut, was used as a substrate for the cultivation of two species of white rot fungi, *Pleurotus eryngii* and *Trametes versicolor*, in the liquid medium. The activities of laccase, lignin, and manganese peroxidases in the growth media were measured to determine their possible involvement in PS microplastic degradation. To assess the influence of *Tenebrio molitor* on PS degradation, as well as the influence of two fungal species on PS microplastic particles, we used Fourier transform infrared spectroscopy (FTIR) to determine whether there were structural changes in the larvae- and fungi-treated EPS and XPS. Finally, UHPLC/Orbitrap MS analysis was performed to identify phenolic compounds as potential products of plastic degradation in fungal media.

We hypothesized that *Tenebrio molitor* consumption of intact EPS and XPS could be an efficient pre-treatment for the formation of MP particles and the initial stage of the change in PS structure (deterioration), which will enable an efficient approach of fungal enzymes to the polymer and cause its further biodegradation. Such an experimental setup could lead to an improved and innovative bio-based process for PS biodegradation.

2. Materials and Methods

2.1. Insect Treatment

The specimens of the insect species *Tenebrio molitor* L. have been grown and maintained in the laboratory of the University of Belgrade Institute for Multidisciplinary Research on the natural food source (wheat flour), and the obtained larvae were used in the experiment. Twenty larvae L5 stage were placed in each of the two plastic boxes with a volume of 5 L, and two types of polystyrene foam were added separately in each box as the sole food source: styrofoam, expanded polystyrene (EPS), and styrodure, extruded polystyrene (XPS) (Figure 1). Both types of polystyrene were in the form of a thick sheet cut into irregular cube-shaped pieces a few centimeters in size (Figure 2). Larvae in two plastic boxes were incubated in the dark at room temperature, around 23–25 °C and 70% humidity for 6 weeks. For the time given, 80% of larvae survived and were transformed into imago, while 20% did not manage to complete the entire life cycle and eventually died. The insect larvae' frass was collected, analyzed, and used as a substrate in experiments with fungi (Figure 2).

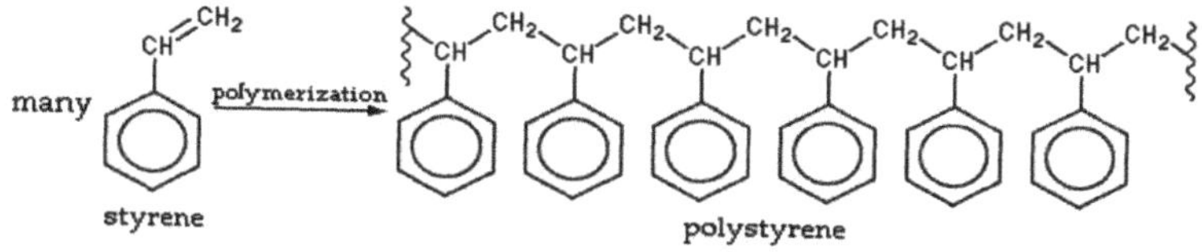

Figure 1. The structure of polystyrene (taken from Yanto et al.) [56].

Figure 2. Larvae of *Tenebrio molitor* feeding on EPS (**left**) and frass of *T. molitor* that was fed on XPS before H_2O_2 treatment (**right**).

2.2. Fungal Treatment

Trametes versicolor (L.) Lloyd strain SerBC was isolated from the fruiting body originating from the forest in northern Serbia, while *Pleurotus eryngii* (DC.) Quél. strain EspCF was received from the mycelia-producing company Cultivos Forestales, Torre de Las Arcas, Spain. The mycelial cultures of *P. eryngii* and *T. versicolor* were maintained in Petri dishes on malt extract agar in the dark at room temperature. To confirm the strains' taxonomic determination, DNA was isolated from mycelia and used for PCR-based amplification with standard primers for fungi (ITS1F and ITS4) and applying standard protocols [57]. The PCR products were sent for Sanger sequencing to Microsyth, Vienna, and the obtained sequences were compared with existing ones in the NCBI database using the BLASTn option from the site [58].

For the experiment on MP degradation by both fungal species, the liquid medium containing 15 g/L of malt extract and 5 g/L of peptone was used. In 100 mL Erlenmeyer flasks, 50 mL of liquid medium was added and sterilized in an autoclave at 114 °C for 30 min. Liquid media were inoculated with 5 one-week-old mycelial plugs of both fungi species separately per flask. Insect larval frass was treated with 30% hydrogen peroxide to decompose organic matter and bacteria from the larvae's intestinal tract, carefully washed with sterile water, dried, and added to the liquid cultures of both fungal species, 300 mg per Erlenmeyer flask. Frass of insect larvae, grown on two types of polystyrene, was added separately. In addition to the experimental treatments, appropriate controls were prepared: liquid medium with added insect larval frass that was not inoculated with fungal mycelia and liquid fungal cultures of fungi to which no plastic substrate was added. All the experimental treatments were incubated at 30 °C in the dark for 30 days.

2.3. Sample Preparation

Control of EPS and XPS, as well as experimental insect frass treated with 30% hydrogen peroxide, were analyzed directly using Fourier transform infrared spectroscopy (FTIR).

The fungi-treated sample preparation for analysis involved separating the liquid and solid phases on filter paper using a vacuum pump. Solid phases, containing fungal mycelia and polystyrene particles, were lyophilized and stored in a freezer at −20 °C until FTIR analysis. Liquid phases in the amount of 4185 g were centrifuged for 30 min. A total of 3 mL of supernatant was collected and used for the analysis of the ligninolytic enzyme activity.

The rest was lyophilized, stored in a freezer at −20 °C, and used for high-performance liquid chromatography (HPLC) analysis.

2.4. Laccase Activity

The determination of laccase activity was carried out according to the protocol of [59], which is based on the oxidation of guaiacol. The reaction mixture of 10 mM sodium acetate buffer, pH 5, 2 mM guaiacol, and the enzyme source was made. After incubation of the reaction mixture at 30 °C for 15 min, the absorbance of oxidized guaiacol was measured at 470 nm (2501 PC Shimadzu, Kyoto, Japan). For the calculation of enzyme activity (U/mL), the extinction coefficient of guaiacol was taken to be ε = 6.74 L/(mmol × cm) as reported by Hosoya [60].

2.5. Mn Peroxidase Activity

Determination of Mn peroxidase activity was performed according to the previously published protocol of Cascielo et al. [61], which is based on ABTS oxidation. The reaction mixture of 0.5 mM ABTS and 0.16 mM $MnCl_2$ in a 40 mM sodium citrate buffer, pH 4.5, was prepared, and the reaction was initiated with the addition of 0.05 mM H_2O_2. The absorbance was measured at 420 nm (2501 PC Shimadzu, Kyoto, Japan) against distilled water as the reference mixture (ε = 36,000 $M^{-1}cm^{-1}$). For the enzyme activity calculation, it was assumed that one unit of enzyme activity (U) corresponds to the amount of enzyme needed to oxidize 1 µmol of ABTS per minute at 25 °C.

2.6. Lignin Peroxidase Activity

The determination of lignin peroxidase activity was performed according to the previously published protocol [62], which is based on the oxidation of Azure B dye. The reaction mixture, consisting of a 125 mM sodium tartrate buffer (pH 3.0), 0.16 mM azure B, and enzyme filtrate in a total volume of 1 mL, was initiated with the addition of 2 mM H_2O_2. The decrease in absorbance was measured at 651 nm (ε = 48.8 L/mmol·cm; 2501 PC Shimadzu, Kyoto, Japan). One unit of enzyme activity (U) was expressed as a ΔA of 0.1 unit per minute per ml of the culture filtrate.

2.7. UHPLC/Orbitrap MS Analysis

The phenolic compounds were identified in methanol extracts using an LC/MS (Thermo Scientific™ Vanquish™ Core HPLC system, Waltham, MA, USA) coupled to the Orbitrap Exploris 120 mass spectrometer (San Jose, CA, USA). All LC/MS parameters and procedures for data acquisition, processing, and analysis are explained in detail in Stojkovic et al. [63] and Popović et al. [64]. Lyophilized samples were used for HPLC analysis; they were dissolved in 100% methanol, shaken at 120 rpm for 4 h, after which the supernatant was filtered through 22 µm nylon syringe filters and used undiluted for analysis. Phenolic compounds were tentatively identified based on their chromatographic behavior and mass spectrometric (MS and MS^2) data, through comparison with available standards and relevant literature sources. A total of 20 compounds were identified and quantified by LC-MS in all the extracts and examined in heatmaps.

2.8. FTIR Analysis

ATR-FTIR spectroscopy (SpectrumTwo, Perkin Elmer, Waltham, MA, USA) was used to analyze the samples: control samples of expanded and extruded polystyrene, frass of experimental insects, and samples treated with fungi, which contained fungal mycelia and polystyrene particles and fungal mycelia. The sample preparation procedure is described in detail in Section 2.3. Spectra were recorded in the range of 4000 to 400 cm^{-1}, with

a resolution of 8 cm^{-1}. For each sample, the full FTIR spectrum was recorded with 16 accumulations, and the analysis was performed in three replicates.

2.9. Statistical Analyses

Two-factor ANOVA with a balanced design (n = 3) was used to examine the differences between the mean values of enzyme activities between individual treatments for the two fungal species and between the two fungal species within individual treatments.

To assess and visualize the similarity between FTIR spectral profiles of fungi cultivated on EPS and XPS, principal component analysis (PCA) was employed as a multivariate technique for dimensionality reduction and pattern recognition. FTIR spectra were initially recorded in the transmittance mode and subsequently transformed into absorbance to enhance interpretability and ensure linearity concerning the concentration, according to the Beer–Lambert law. For the analysis, two specific spectral regions were selected: 400–1800 cm^{-1} and 2700–3700 cm^{-1}. The intermediate region (1800–2700 cm^{-1}) was excluded due to a lack of significant spectral variability, rendering it uninformative for multivariate analysis. Before PCA, spectral pre-processing was applied to improve data quality and comparability. Baseline correction was performed using the rubber band method to minimize baseline drift.

PCA was conducted on the pre-processed spectra, where principal components were computed as linear combinations of the original spectral variables. The eigenvectors (loadings) were derived to maximize the captured variance in the dataset, with each subsequent component orthogonal to the preceding ones. The first two principal components (PC1 and PC2), explaining the largest portion of total variance, were used to visualize the distribution and grouping of samples in a two-dimensional space. All PCA computations were performed using the *Orange Data Mining Toolbox for Data Analysis*, version 3.35 [65]. Graphical representations of PCA results were generated using Microsoft Excel (Microsoft Corporation, One Microsoft Way, Redmond, WA, USA, Version 365).

3. Results

The majority of *Tenebrio molitor* larvae successfully survived until the end of the experiment (some transforming into the imago stadium), feeding on two types of polystyrene foam as the only food source. Larvae were allowed to feed on polystyrene for 45 days, which was enough to collect sufficient amounts of frass with incompletely decomposed polystyrene for the analysis and setting up experiments with white rot fungi.

Fungal strains' taxonomical determination was confirmed by the analysis of the ITS regions of their DNA. The amplified ITS regions sequences revealed 99–100% identity with different sequences of *Trametes versicolor* and *Pleurotus eryngii*, respectively, and were deposited in the NCBI database (accession numbers PV682683 and PV682684).

3.1. Ligninolytic Enzymes Activity Analysis

Mycelial growth was successful and abundant across all experimental treatments. The activity of the examined ligninolytic enzymes was not detected in controls containing polystyrene, making it clear that the detected enzyme activity originates from fungi. The measured values of enzyme activity are shown in Figure 3. High levels and similar patterns of laccase activity were recorded in the media of both strains of white rot fungi used in the experiment. The significantly lower Lac activity was registered in the treatment without a substrate compared to EPS and XPS, but there was no significant difference between the polystyrene treatments for both species (Figure 3a). Both species had high MnP activity values, but activities of the MnP of *T. versicolor* were significantly higher in XPS treatment

than with *P. eryngii* (Figure 3b) and, especially, in the control samples compared to *P. eryngii*. Activities of both enzymes in the control with *T. versicolor* were significantly higher than in the control with *P. eringii*. LiP activity was detected only in the species *T. versicolor* in the absence of substrate and the treatment with EPS, but there were no significant differences between these two treatments.

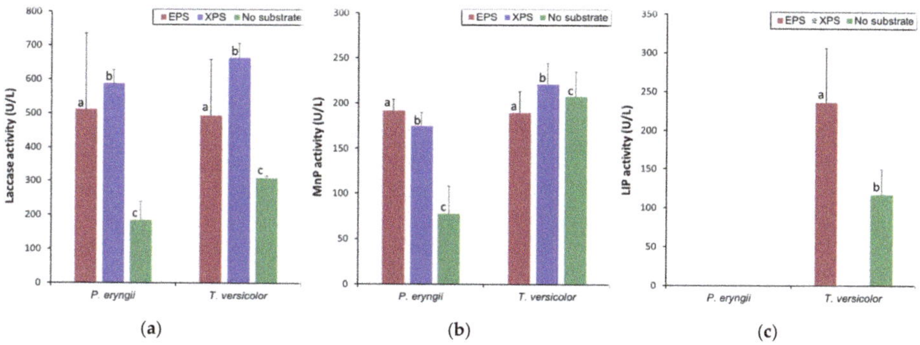

Figure 3. The enzyme activities measured in the media of *Pleurotus eryngii* and *Trametes versicolor* ((**a**) laccase activity in fungal mycelia; (**b**) Mn-peroxidase activity in fungal mycelia; (**c**) lignin peroxidase activity in fungal mycelia; EPS—expanded polystyrene, styrofoam; XPS—extruded polystyrene, styrodure). a, b, c are statistical classes indicating statistical significance of the differences between enzyme activities.

3.2. UHPLC/Orbitrap MS Analyses of the Potential Products of Plastic Degradation

Detection frequencies of phenolic compounds in the control (T.v.C for *T. versicolor*, P.e.C for *P. eryngii*) and treatments with XPS, styrodure (T.v.XPS/P.e.XPS), or EPS, styrofoam (T.v.EPS/P.e.EPS), are summarized in heat maps (Figures 4 and 5).

Phenolic substances were detected in all treatments, but their profiles in the media with different polystyrene forms added expressed certain differences (Figures 4 and 5). In the control mediums with added EPS or XPS only, small amounts of phenolic substances were detected sporadically. Ferulic and protocatechuic acids, daidzein, apigenin, and isorhamnetin were present in the media with ESP, while in media with XPS, apigenin was absent, and additionally, hydroxybenzoic, 5-O-caffeoylquinic acids, and quercetin were recorded. Phenolic profiles of the media of the two fungal species differ significantly (Figures 4 and 5). Caffeic and p- p-hydroxybenzoic acids, naringenin, luteolin, and isorhamnetin were strongly produced in every treatment with *T. versicolor*, while protocatechuic and humic acids and daidzein were characteristically strongly expressed in all treatments with *P. eryngii*. Syringic acid and quercetin appeared only in media with polystyrene added to fungal mycelia, while kaempferol-3-O-glucoside appeared only in media with EPS added to mycelia of *P. eryngii*. Other phenolics were detected in the media in varying patterns (Figures 4 and 5).

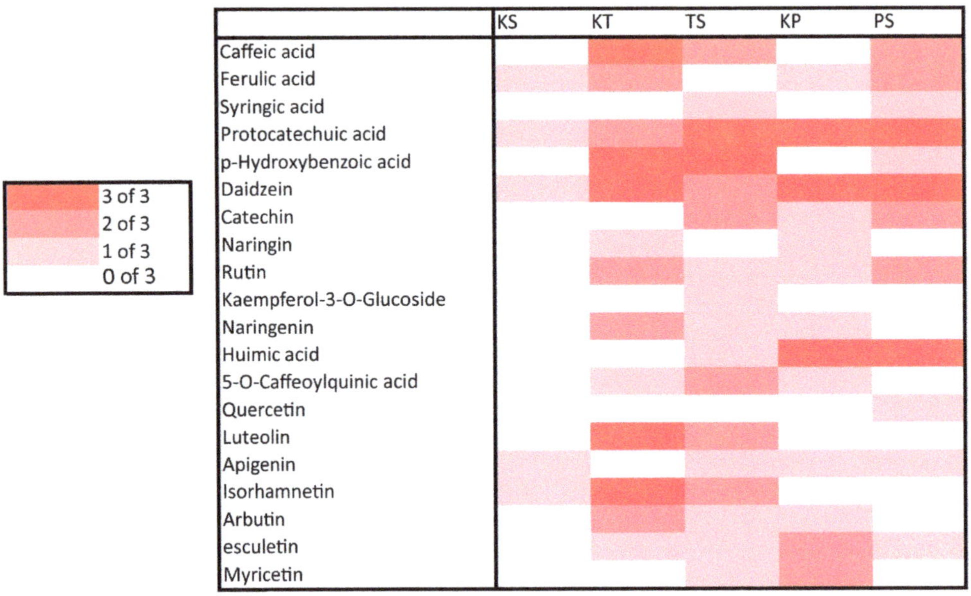

Figure 4. Detection frequency of phenolic compounds in methanolic extracts of *T. versicolor* and *P. eryngii* cultivated on larval frass derived from XPS. Detection was based on LC-MS analysis across three biological replicates per condition. Darker shades indicate consistent compound presence (3/3), while lighter shades and white reflect variable or absent detection.

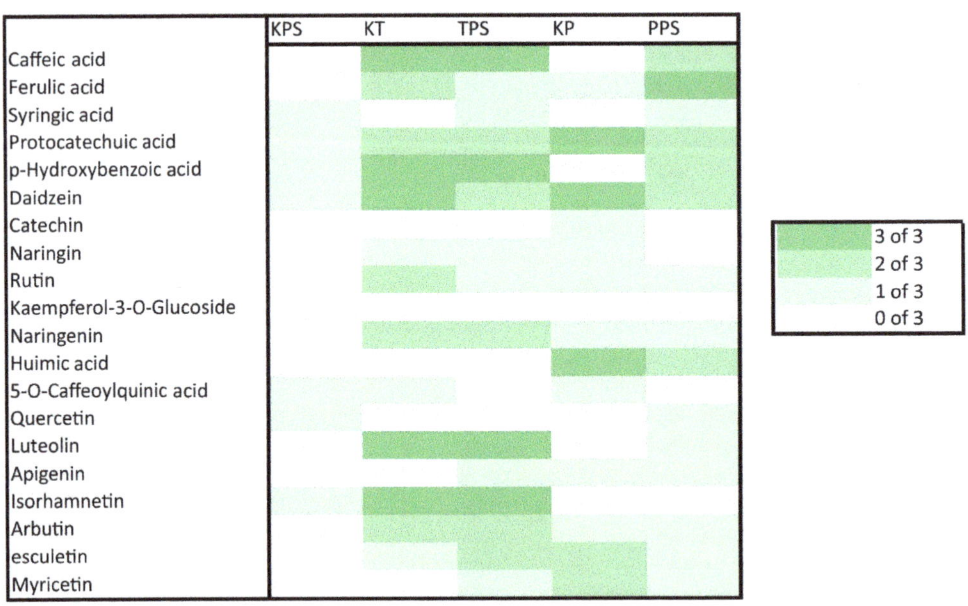

Figure 5. Detection frequency of phenolic compounds in methanolic extracts of *T. versicolor* and *P. eryngii* cultivated on larval frass derived from EPS. Detection was based on LC-MS analysis across three biological replicates per condition. Darker shades indicate consistent compound presence (3/3), while lighter shades and white reflect variable or absent detection.

3.3. FTIR Analysis of the EPS/XPS Exposed to the *T. molitor*, *P. eryngii*, and *T. versicolor* in the Experimental Treatments

The FTIR profiles of both types of PS used in the experiment, expanded and extruded, look almost identical, except for minor differences in the intensity of a few individual peaks. The observed absorption bands in the FTIR spectrum and band assignment of polystyrene controls are shown in Table 1.

Table 1. Observed absorption bands in the FTIR spectrum and band assignment of polystyrene controls.

Wave Number (cm^{-1})	Band Assignment	References
3082	Aromatic C–H stretching vibrations	Olmos et al. [66]
3060	Aromatic C–H stretching vibrations	Olmos et al. [66]
3026	Aromatic C–H stretching vibrations	Olmos et al. [66]
2917	C-H bond stretching	Meenashi et al. [67]
2848	Symmetric stretching vibrations of methylene groups –CH_2	Olmos et al. [66]
1601	C-C stretching vibrations in the aromatic ring	Olmos et al. [66]
1493	C-C stretching vibrations in the aromatic ring	Olmos et al. [66]
1452	C–H deformation vibration of CH_2	Bhuto et al. [68]
1028	The in-plane C–H bending of the phenyl ring	Olmos et al. [66]
907	Out-of-plane C–H bending of the phenyl ring	Olmos et al. [66]
755	Out-of-plane C–H bending of the phenyl ring	Ilijin et al. [69]
696	Out-of-plane C–H bending of the phenyl ring	Ilijin et al. [69]

The FTIR spectrum of larval frass derived from expanded polystyrene and presented in Figure 6 (T.m. EPS) had a similar profile to the untreated polystyrene control. In addition to minor changes in the intensity of the peaks, one new peak that was not present in the control FTIR spectrum can be observed at 1647 cm^{-1} (Figure 6). In the FTIR spectra after incubation with two fungal species, significant differences can be observed, both concerning the control and polystyrene after passing through the digestive system of insects and to the two fungal species. After incubation with fungi *P. eryngii* (Figure 6, P.e.EPS), a loss of peaks at 3082, 3060, 3026, 1493, and 1452 cm^{-1}, as well as the emergence of new peaks at 3272, 1634, 1416, 1310, and 558 cm^{-1} can be observed in the FTIR spectrum. Shifts in absorption peaks were evinced from 2917 to 2921 cm^{-1}, 2848 to 2853 cm^{-1}, 1028 to 1035 cm^{-1}, 907 to 890 cm^{-1}, and 538 to 531 cm^{-1}. After incubation with *T. versicolor* (Figure 6, T.v.EPS), new peaks can be observed in the FTIR spectrum at 3256 and 1306 cm^{-1} and a shift in absorption peaks from 2917 to 2919 cm^{-1}, 2848 to 2851 cm^{-1}, and 1028 to 1029 cm^{-1}.

The PCA-based visualization of the relationships between the different spectra recorded by FTIR spectra of EPS biodegradation is presented in Figure 7. In this analysis, we introduced spectra of the fungal mycelia (Figure S1) to distinguish signals of fungi from those of plastic polymers (Figure S2). The clear differentiation of the spectra according to the treatment is visible. Spectra of *T. molitor*-treated EPS were the most similar to the EPS control, while spectra of fungal mycelia strongly differed from those of the EPS containing materials (differentiated with PC1, Figure 7).

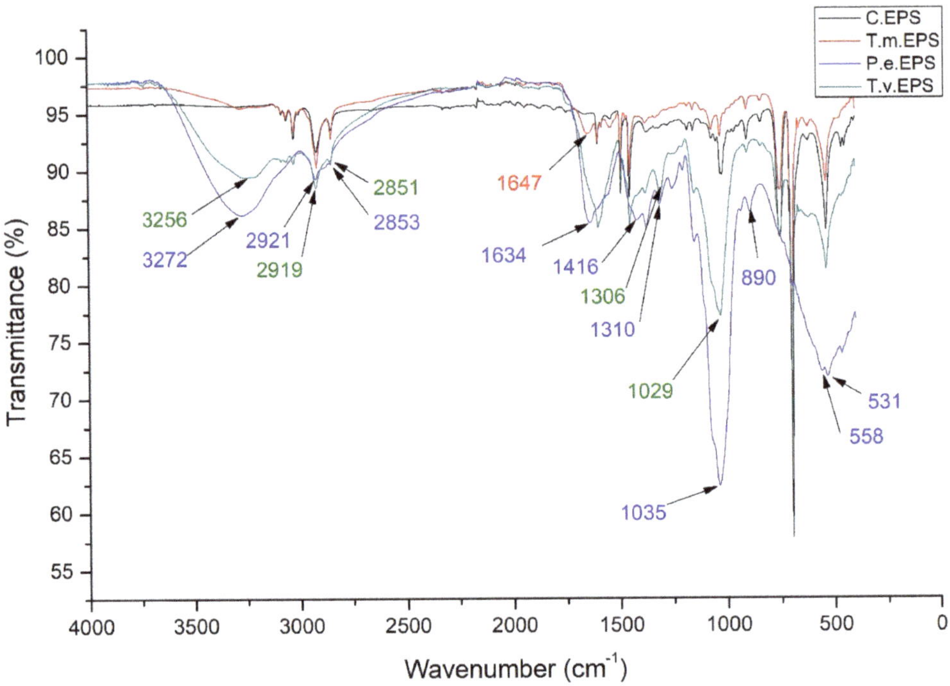

Figure 6. Average FTIR spectra of experimental treatments with expanded polystyrene (C.EPS—expanded polystyrene control; T.m.EPS—polystyrene after the digestive system of *T. molitor* larvae; P.e.EPS—polystyrene in insect frass treated with *P. eryngii*; T.v.EPS—polystyrene in insect frass treated with *T. versicolor*).

The FTIR spectrum of larval frass derived from extruded polystyrene is presented in Figure 8.

Compared to the control, the largest differences in the obtained FTIR spectra of EPS after passing through the digestive system of insect larvae are the appearance of new peaks at 3275, 1637, 1114, and 472 cm^{-1} (Figure 8, T.m.XPS). In the incubation treatment with the species *P. eryngii* (Figure 7, P.e.XPS), the loss of peaks was observed at 3026, 1493, and 1452 cm^{-1}, the appearance of new peaks at 3272, 1634, 1416, 1310, 558, and 464 cm^{-1}, as well as the shift of peaks from 2917 to 2921 cm^{-1}, 2848 to 2853 cm^{-1}, 1028 to 1035 cm^{-1}, 907 to 890 cm^{-1} and 538 to 531 cm^{-1}, similar to the treatment with expanded polystyrene. In the experimental treatment with the species *T. versicolor* (Figure 8, P.e.XPS), the following changes were observed in the FTIR spectrum of extruded polystyrene: the emergence of new peaks at 3256 and 1306 cm^{-1} and shift in absorption peaks from 2917 to 2919 cm^{-1}, 2848 to 2851 cm^{-1}, and 1028 to 1029 cm^{-1}, as also detected in the treatment with expanded polystyrene (Figure 8).

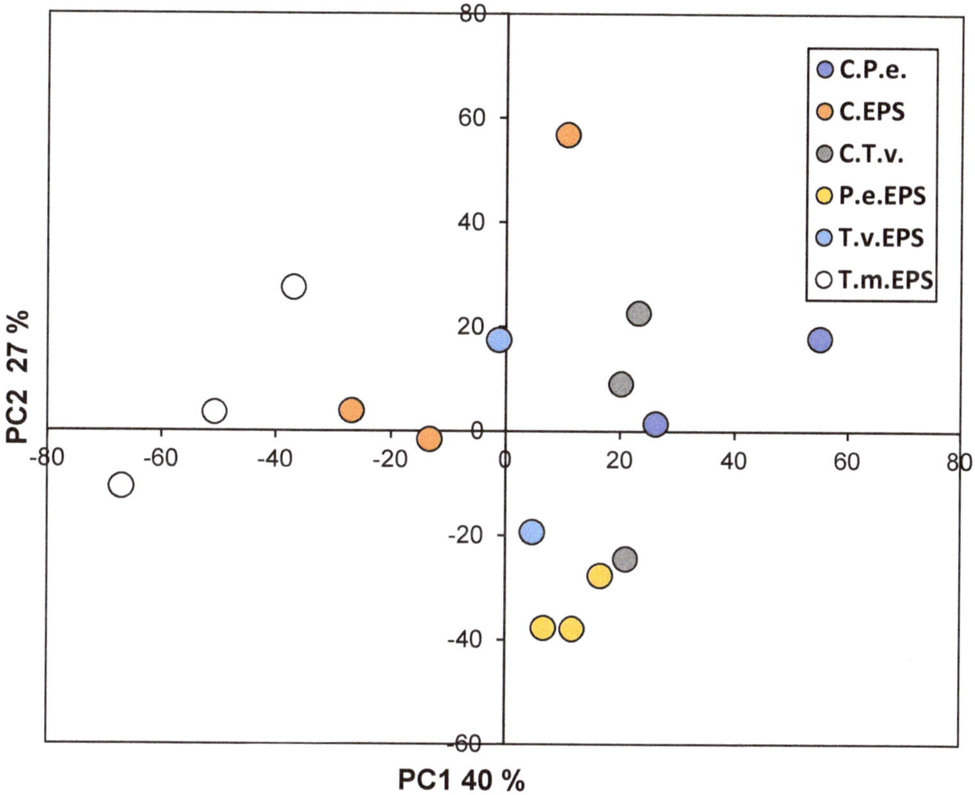

Figure 7. The score plot of the first two principal components (PC1 and PC2) obtained from PCA of FTIR spectra of treatments based on EPS The plot shows treatment distribution based on spectral similarity in the regions 400–1800 cm^{-1} and 2800–3700 cm^{-1}. PC1 and PC2 together explain 83% of the total variance. (C.EPS—expanded polystyrene control; T.m.EPS—polystyrene after the digestive system of *T. molitor* larvae; P.e.EPS—polystyrene in insect frass treated with *P. eryngii*; T.v.EPS—polystyrene in insect frass treated with *T. versicolor*; C.P.e.—control of mycelia of *Pleurotus eryngii*; C.T.v.—control of mycelia of *Trametes versicolor*).

The PCA-based visualization of the relationships between the different spectra recorded by FTIR spectra of XPS biodegradation is presented in Figure 9. The spectra analyses showed even better differentiation of treatments with different biological agents along PC1, with PS treatments on one and fungal mycelia on the other side of the graph (Figure 9). It is also very visible that XPS degradation of T. molitor produced quite different outcomes than fungal treatments.

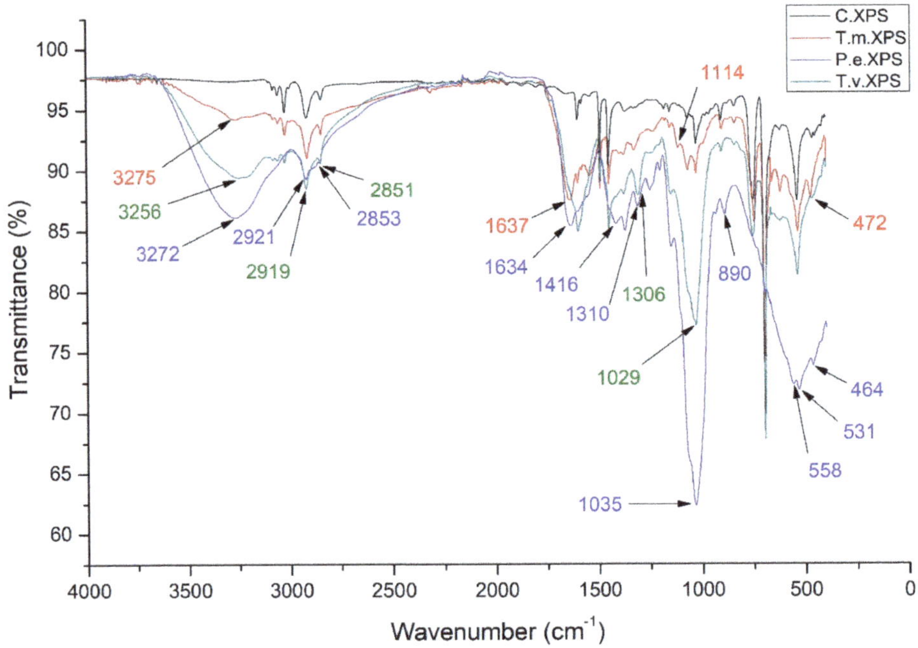

Figure 8. Average FTIR spectra of experimental treatments with extruded polystyrene (C.XPS—untreated extruded polystyrene control; T.m.XPS—polystyrene after the digestive system of *T. molitor* larvae; P.e.XPS—polystyrene in insect frass treated with *P. eryngii*; T.v.XPS—polystyrene in insect frass treated with *T. versicolor*).

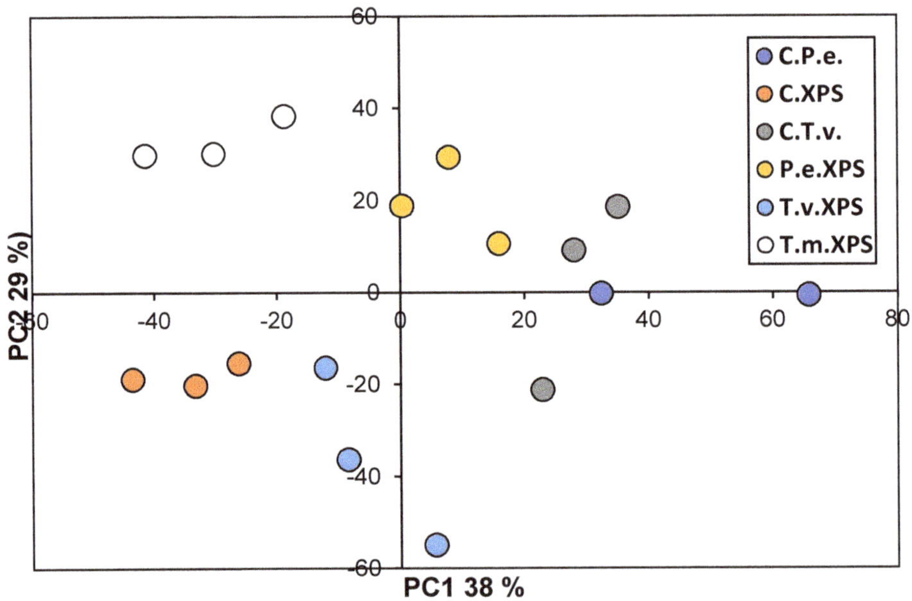

Figure 9. The score plot of the first two principal components (PC1 and PC2) obtained from PCA of FTIR spectra of treatments based on XPS The plot shows treatment distribution based on spectral

similarity in the regions 400–1800 cm^{-1} and 2800–3700 cm^{-1}. PC1 and PC2 together explain 83% of the total variance. (C.EPS—expanded polystyrene control; T.m.EPS—polystyrene after the digestive system of *T. molitor* larvae; P.e.EPS—polystyrene in insect frass treated with *P. eryngii*; T.v.EPS—polystyrene in insect frass treated with *T. versicolor*; C.P.e.—control of mycelia of *Pleurotus eryngii*; C.T.v.—control of mycelia of *Trametes versicolor*).

4. Discussion

In this contribution, we demonstrated that *Tenebrio molitor* larvae could feed on polystyrene materials, destroying their structure and transforming them into small particles (MPs) within their digestive tract. These particles still consisted of polystyrene molecules, but the signs of the polymer's destabilization were evident. When offered to the two white rot fungal strains as a substrate, these particles were further degraded by the activities of laccases and Mn peroxidases produced by both fungi, resulting in the excretion of the various phenolic substances in the growth media.

T. molitor belongs to the order Coleoptera, family *Tenebrionidae*, and is a cosmopolitan species. Previous research indicated that polystyrene is susceptible to degradation in the intestinal tract of *T. molitor* larvae [70]. When feeding on polystyrene, the larvae nibble and chew PS, thereby mechanically breaking it down, which can be accepted as biodeterioration—the first step in its biodegradation. Further biodegradation takes place in the digestive tract of larvae through gut-microbe-dependent oxidative digestive machinery [70]. Among the gut microbes, *Enterococcus* and *Enterobacteriaceae* play a crucial role in this polystyrene biodegradation process [71]. In previously published papers, it was claimed that the development of *T. molitor* larvae to adulthood is impossible due to a lack of various nutrients, primarily nitrogen, but in our experiments, majority of larvae successfully survived feeding on polystyrene as its sole food source until the end of the experiment (45 days) and even completed their development into adult beetles. The research conducted by Peng et al. [70] indicates that the successful survival of *T. molitor* larvae feeding only on polystyrene is possible for about 4–5 weeks, which is similar to what we observed here.

It is well known that white rot fungi, like those used in experiments in this contribution, produce ligninolytic enzymes that are characterized by low substrate specificity [25,26]. High Lac activity was detected in the experiment with both strains of fungi and with both types of larval frass (Figure 3a). Due to a relatively low redox potential [72], laccases can directly oxidize only some phenolic compounds, but the spectrum of their activity is significantly expanded through redox mediators. These are phenolic compounds of low molecular weight, such as hydroxycinnamic acids [73]. Some of the compounds of this group, caffeic and ferulic acids, have been detected via UHPLC/Orbitrap MS analysis (Figures 4 and 5). Since the enzymes as well as phenolic substances were present in the control treatment with fungi without the addition of PS, it can be assumed that both fungi may produce Lac in the presence of growth factors in the media and accelerate the production of such mediators. Additionally, both fungi also exhibited high Mn peroxidase activity with both types of larval frass (Figure 3b). MnP can directly oxidize phenolic compounds [74] but cannot directly oxidize non-phenolic aromatic structures of lignin [75]. Mn^{3+} cations, which are formed as a product of the catalytic activity of Mn peroxidases, are stabilized by forming chelates with various organic acids, such as oxalic acid, which are derivatives of fungal metabolism. These chelates of low molecular weight function as highly reactive redox mediators that non-specifically attack and oxidize different polymers [76]. Both Lac and MnP cause similar reactions when exposed to artificial polymers—they attack them by causing redox reactions, which produce ROS, and then oxidize functional groups

within the polymer, thereby altering their structure to a more destructible state [77]. Further exposure of the polymer to these enzymes and generated ROS will cause its fragmentation, generating smaller structures with polar functional groups that are more reactive with enzymes [34]. Small products of such degradation can be transported into the cells for further metabolism [78].

LiP is an enzyme that has a high redox potential and can oxidize non-phenolic and phenolic aromatic compounds [79]. A typical substrate of lignin peroxidase is veratryl alcohol, which is produced as a derivative of the secondary metabolism of fungi [80]. This compound, through the mechanism of free radical production, can serve as a mediator of the oxidation of various compounds that are otherwise not substrates of lignin peroxidase. Fungi from the genus *Pleurotus* sp. do not produce typical lignin peroxidase, which was confirmed in our experiments (Figure 3c); instead, they produce versatile peroxidases [81]. LiP was detected in the *T. versicolor* medium in the absence of substrate and the treatment with EPS but, surprisingly, not in the treatment with larval frass derived from XPS. It could be possible that the firm structure of XPS or some additives blocked the production of LiP by *T. versicolor* (Figure 3c)

Obtained UHPLC/Orbitrap MS analysis data reveal that XSP frass strongly stimulates the biosynthesis of key hydroxycinnamic acids—caffeic and ferulic—in *P. eryngii*, whereas controls show no such induction. The selective emergence of syringic acid under ESP and XSP treatments (Figures 4 and 5) suggests PS induced radical-scavenging pathways. Protocatechuic and hydroxybenzoic acids' consistent detection across all *T. versicolor* conditions suggests a basal role in reactive-oxygen-species detoxification [82]. Flavonoid profiling indicates that quercetin production was detected in the control as well as in the media of *P. eryngii* with (XPS), confirming its role as a redox modulator included in PS degradation. Naringin moderate stimulation with EPS across treatments reflects activation of specific biosynthetic branches. The presence of EPS induced kaempferol-3-O-glucoside in *T. versicolor* treatment points to interspecies variation in flavonol glycosylation [83]. Finally, the detection of daidzein, apigenin, and esculetin in all fungal treatments suggests activation of fungi-specific metabolic pathways. Specific induction of chlorogenic acid across treatments emphasizes its species-specific antioxidant function, whereas the detection of arbutin and humic acid in species species-specific manner confirms their probable negligible role in plastic degradation. Overall, *T. versicolor* demonstrates a more robust induction of core antioxidant phenolics in response to XSP frass, suggesting these compounds as biomarkers of successful PS and XSP degradation and highlighting opportunities for optimizing fungal biodegradation through strain selection or engineering. The phenolic profiles correspond to the enzymatic profiles, confirming species-specific pathways for degradation of aromatic polymers.

The obtained FTIR spectra of larval frass derived from EPS were similar to the spectra of polystyrene in the control, while minor differences in peak intensity and the appearance of one new peak at 1647 cm^{-1} were detected. The FTIR spectrum of larval frass derived from extruded polystyrene is similar, with several new peaks at 3275, 1637, 1114, and 472 cm^{-1}. Peng et al. [70] indicated that the peak at 1650 cm^{-1} represents −C=C− stretch and detected peaks at 1637 and 1647 cm^{-1}, probably representing the same functional group. The peak at 3275 cm^{-1} is part of the 2500–3500 cm^{-1} area that is associated with the hydrogen bond of hydroxyl groups and/or carboxylic acid groups and suggests a shift from hydrophobic to more hydrophilic surface properties [84]. The incorporation of oxygen functional groups is considered one of the first steps and key evidence of biodegradation of polystyrene polymers [84]. According to Olmos et al. [66], bands in the region from 3200 to 2800 cm^{-1} can be assigned to C–H stretching vibrations. Bands at 3082, 3060, and

3026 cm^{-1} correspond to absorptions from the C–H stretching vibrations in benzene rings, while the bands at 2917 and 2948 cm^{-1} come from asymmetric and symmetric stretching vibrations from CH_2 groups in the main polystyrene chain. The absorptions at 1601 and 1493 cm^{-1} come from C–C stretching vibrations in the aromatic rings. According to Bhutto et al. [67], band at 1452 cm^{-1} can be assigned to C–H deformation vibration of CH_2 groups. According to Ilijin et al. [69], the bands in the regions from 1066 to 1028 cm^{-1} and from 755 to 696 cm^{-1} can be assigned to C–H bending vibrations of the ring in plane and C–H out-of-plane bending vibrations of the benzene ring. The main absorption peaks in these regions are observed at 1028, 907, 755, and 696 cm^{-1}.

Principal component analysis (PCA) score plots based on FTIR absorbance spectra of fungal samples grown on polystirene were able to distinguish clearly between the polymer structures of different treatments (Figures 7 and 9). As expected, samples from the C.P.e. and C.T.v.—the spectra of fungal mycelia exhibited the highest scores along PC1, indicating pronounced spectral differences from the other treatments, which are more closely clustered along this axis. Interestingly, spectra of *T. molitor* treatment exhibited the lowest scores along PC1, including C.EPS, indicating a very different mechanism of PS destruction in comparison to fungal treatments (Figure 7). Fungal treatments, on the other hand, differed significantly from the C.EPS, as well as from fungal controls (Figure 7). The pattern in the distribution of spectral profiles for the control and treatments with extruded polystyrene (XPS) showed similarities to that observed in the case of EPS treatments. However, PC2 made a clear differentiation between T.m.XPS and C.XPS, as well as between P.e.XSP and T.v.XSP. The observed lack of LP activity in *T. versicolor* treatment with XSP (Figure 3c) may be an explanation of such differences and the proximity of T.v.XSP spectra to C.XSP, as well as a difference between the spectra of the two fungal treatments.

In general, PCA of the FTIR spectra of different treatments revealed very clear differentiation between spectra produced by the PS molecule after *T. molitor* deterioration of original PS products into PS microplastics, fungal mycelia, and treated EPS and XPS (Figures 7 and 9). Comparing the spectra of PS material before *T. molitor* treatment (Figure S1) and mycelia (Figure S2), it is easy to spot the differences between spectra that are visible in Figures 6 and 8. These graphs are clear evidence that *T. molitor*, as well as both fungal strains, can degrade PS, but in species species-specific manner.

5. Conclusions

PS of different shapes can deteriorate and transform into the particle size of microplastics by *Tenebrio molitor* larvae in a relatively short time (4–6 weeks). Such particles can be degraded by white rot fungi through their enzymatic systems used for lignin degradation, like Lac, MnP, and LiP. These enzymes are known to form films on the polymer surface, eroding it through the oxidation of external functional groups and creating further ROS that attack the polymers. These new formations are susceptible to further action by oxygenases in a chain reaction. Thus far, this process has not led to the successful degradation of entire polymers, but most laboratory experiments, including ours, likely do not last long enough to enable such outcomes. For example, experiments lasting 50 days revealed stronger PS degradation by other white rot fungi [85]. It can be hypothesized that adding certain molecule accelerators, like those produced in our fungal media (caffeic and ferulic acids, coniferyl alcohol, etc), may speed up the process. Our results indicate that *T. versicolor* and *P. eryngii* employ similar, but not identical, enzymatic mechanisms, suggesting that including more than one fungal species in the process may enhance its efficiency. Even though we present very early-stage results of the combination of different organisms in the biodegradation of PS, optimizing this process could lead to complete PS degradation.

Supplementary Materials: The following supporting information can be downloaded at: https://www.mdpi.com/article/10.3390/polym17131772/s1, Figure S1: FTIR spectra of fungal mycelia (CP.e—*P. eryngii;* CT.v—*T. versicolor*); Figure S2: FTIR spectra of expanded (CEPS) and extruded polystyrene (CXPS).

Author Contributions: Conceptualization, Ž.M., SK. and S.S.; methodology S.K., Ž.M., D.M., M.D. and S.S.; software M.D.; validation, Ž.M. and S.K.; formal analysis, S.S. and K.A.; investigation Ž.M., S.K. and S.S.; resources, Ž.M. and S.K.; data curation S.S. and Ž.M.; writing—original draft preparation, S.S. and K.A.; writing—review and editing, Ž.M. and S.S.; supervision, Ž.M. and S.K.; project administration, Ž.M. and S.K.; funding acquisition, Ž.M. and S.K. All authors have read and agreed to the published version of the manuscript.

Funding: This research was supported by the Science Fund of the Republic of Serbia, grant No 7742318, "Evaluation of the Microplastics in the Soils of Serbia"—EMIPLAST S.o.S. and by the Ministry of Science, Technological Development and Innovations of the Republic of Serbia, grant No 451-03-136/2025-03/200053.

Institutional Review Board Statement: Not applicable.

Data Availability Statement: All data are available in the manuscript or Supplementary Data.

Conflicts of Interest: The authors declare no conflicts of interest.

Abbreviations

The following abbreviations are used in this manuscript:

PS	Polystyrene
ESP	Expanded Polystyrene
XPS	Extruded Polystyrene
Lac	Laccase
MnP	Manganese Peroxidase
LiP	Lignin Peroxidase

References

1. Srikanth, M.; Sandeep, T.S.R.S.; Sucharitha, K.; Godi, S. Biodegradation of plastic polymers by fungi: A brief review. *Bioresour. Bioprocess.* **2022**, *9*, 42. [CrossRef] [PubMed]
2. Hidalgo-Crespo, J.; Moreira, C.M.; Jervis, F.X.; Soto, M.; Amaya, J.L.; Banguera, L. Circular economy of expanded polystyrene container production: Environmental benefits of household waste recycling considering renewable energies. *Energy Rep.* **2022**, *8*, 306–311. [CrossRef]
3. Temporiti, M.E.E.; Nicola, L.; Nielsen, E.; Tosi, S. Fungal Enzymes Involved in Plastics Biodegradation. *Microorganisms* **2022**, *10*, 1180. [CrossRef] [PubMed]
4. Ho, B.T.; Roberts, T.K.; Lucas, S. An overview on biodegradation of polystyrene and modified polystyrene: The microbial approach. *Crit. Rev. Biotechnol.* **2018**, *38*, 308–320. [CrossRef]
5. Othman, A.R.; Hasan, H.A.; Muhamad, M.H.; Ismail, N.I.; Abdullah, S.R.S. Microbial degradation of microplastics by enzymatic processes: A review. *Environ. Chem. Lett.* **2021**, *19*, 3057–3073. [CrossRef]
6. Wu, W.M.; Criddle, C.S. Characterization of biodegradation of plastics in insect larvae. *Methods Enzymol.* **2021**, *648*, 95–120.
7. Chiellini, E.; Solaro, R. Biodegradable polymeric materials. *Adv. Mater.* **1996**, *8*, 305–313. [CrossRef]
8. Shah, A.A.; Hasan, F.; Hameed, A.; Ahmed, S. Biological degradation of plastics: A comprehensive review. *Biotechnol. Adv.* **2008**, *26*, 246–265. [CrossRef]
9. Wu, F.; Guo, Z.; Cui, K.; Dong, D.; Yang, X.; Li, J.; Wu, Z.; Li, L.; Dai, Y.; Pan, T. Insights into characteristics of white rot fungus during environmental plastics adhesion and degradation mechanism of plastics. *J. Hazard. Mater.* **2023**, *448*, 130878. [CrossRef]
10. Zara, Z.; Mishra, D.; Pandey, S.K.; Csefalvay, E.; Fadaei, F.; Minofar, B.; Řeha, D. Surface interaction of ionic liquids: Stabilization of polyethylene terephthalate-degrading enzymes in solution. *Molecules* **2022**, *27*, 119. [CrossRef]

11. Wallace, N.E.; Adams, M.C.; Chafin, A.C.; Jones, D.D.; Tsui, C.L.; Gruber, T.D. The highly crystalline PET found in plastic water bottles does not support the growth of the PETase-producing bacterium *Ideonella sakaiensis*. *Environ. Microbiol. Rep.* **2020**, *12*, 578–582. [CrossRef] [PubMed]
12. Tokiwa, Y.; Calabia, B.P.; Ugwu, C.U.; Aiba, S. Biodegradability of plastics. *Int. J. Mol. Sci.* **2009**, *10*, 3722–3742. [CrossRef] [PubMed]
13. Bucher, V.V.C.; Pointing, S.B.; Hyde, K.D.; Reddy, C.A. Production of wood decay enzymes, loss of mass, and lignin solubilization in wood by diverse tropical freshwater fungi. *Microb. Ecol.* **2004**, *48*, 331–337. [CrossRef]
14. Pointing, S.B.; Parungao, M.M.; Hyde, K.D. Production of wood-decay enzymes, mass loss and lignin solubilization in wood by tropical Xylariaceae. *Mycol. Res.* **2003**, *107*, 231–235. [CrossRef]
15. Boer, W.; Folman, L.B.; Summerbell, R.C.; Boddy, L. Living in a fungal world: Impact of fungi on soil bacterial niche development. *FEMS Microbiol. Rev.* **2005**, *29*, 795–811. [CrossRef]
16. Stefanović, S.; Dragišić Maksimović, J.; Maksimović, V.; Bartolić, D.; Djikanović, D.; Simonović Radosavljević, J.; Mutavdžić, D.; Radotić, K.; Marjanović, Ž. Functional differentiation of two autochthonous cohabiting strains of *Pleurotus ostreatus* and *Cyclocybe aegerita* from Serbia in lignin compound degradation. *Bot. Serb.* **2023**, *47*, 135–143. [CrossRef]
17. Perera, P.; Deraniyagala, A.S.; Sashikala Mahawaththage, M.P.; Herath, H.; Kumari Rajapakse, C.S.; Wijesinghe, P.; Attanayake, R.N. Decaying hardwood associated fungi showing signatures of polyethylene degradation. *BioResources* **2021**, *16*, 7056. [CrossRef]
18. Kirk, T.K.; Farrell, R.L. Enzymatic 'combustion': The microbial degradation of lignin. *Annu. Rev. Microbiol.* **1987**, *41*, 465–501. [CrossRef]
19. Hatakka, A. Lignin-modifying enzymes from selected white-rot fungi: Production and role from in lignin degradation. *FEMS Microbiol. Rev.* **1994**, *13*, 125–135. [CrossRef]
20. Nicolcioiu, M.B.; Popa, G.; Matei, F. Biochemical investigations of different mushroom species for their biotechnological potential. In *Proceedings of the Agriculture for Life, Life for Agriculture*; Academia Press: Bucharest, Romania, 2018; pp. 562–567.
21. Kumar, A.; Chandra, R. Ligninolytic enzymes and its mechanisms for degradation of lignocellulosic waste in environment. *Heliyon* **2020**, *6*, e03170. [CrossRef]
22. Harms, H.; Schlosser, D.; Wick, L.Y. Untapped potential: Exploiting fungi in bioremediation of hazardous chemicals. *Nat. Rev. Microbiol.* **2011**, *9*, 177–192. [CrossRef] [PubMed]
23. Pollegioni, L.; Tonin, F.; Rosini, E. Lignin-degrading enzymes. *FEBS J.* **2015**, *282*, 1190–1213. [CrossRef] [PubMed]
24. Datta, R.; Kelkar, A.; Baraniya, D.; Molaei, A.; Moulick, A.; Meena, R.S.; Formanek, P. Enzymatic degradation of lignin in soil: A review. *Sustainability* **2017**, *9*, 1163. [CrossRef]
25. Conesa, A.; Punt, P.J.; van den Hondel, C.A. Fungal peroxidases: Molecular aspects and applications. *J. Biotechnol.* **2002**, *93*, 143–158. [CrossRef]
26. Maciel, M.J.M.; Ribeiro, H.C.T. Industrial and biotechnological applications of ligninolytic enzymes of the basidiomycota: A review. *Electron. J. Biotechnol.* **2010**, *13*, 14–15.
27. Santo, M.; Weitsman, R.; Sivan, A. The role of the copper-binding enzymelaccase-in the biodegradation of polyethylene by the actinomycete *Rhodococcus ruber*. *Int. Biodeter. Biodegr.* **2013**, *84*, 204–210. [CrossRef]
28. Mate, D.M.; Alcalde, M. Laccase: A multi-purpose biocatalyst at the forefront of biotechnology. *Microb. Biotechnol.* **2016**, *10*, 1457–1467. [CrossRef]
29. Han, M.L.; An, Q.; Ma, K.Y.; An, W.N.; Hao, W.Y.; Liu, M.Y.; Shi, W.Y.; Yang, J.; Bian, L.S. A comparative study on the laccase activity of four Basidiomycete fungi with different lignocellulosic residues via solid-state fermentation. *BioResources* **2021**, *16*, 3017–3031. [CrossRef]
30. Demarche, P.; Junghanns, C.; Nair, R.R.; Agathos, S.N. Harnessing the power of enzymes for environmental stewardship. *Biotechnol. Adv.* **2012**, *30*, 933–953. [CrossRef]
31. Surwase, S.V.; Patil, S.A.; Srinivas, S.; Jadhav, J.P. Interaction of small molecules with fungal laccase: A surface plasmon resonance based study. *Enzym. Microb. Technol.* **2016**, *82*, 110–114. [CrossRef]
32. Desai, S.S.; Nityanand, C. Microbial laccases and their applications: A review. *Asian J. Biotechnol.* **2011**, *3*, 98–124. [CrossRef]
33. Zhao, X.; Huang, X.; Yao, J.; Zhou, Y.; Jia, R. Fungal growth and manganese peroxidase production in a deep tray solid-state bioreactor, and in vitro decolorization of poly R-478 by MnP. *J. Microbiol. Biotechnol.* **2015**, *25*, 803–813. [CrossRef] [PubMed]
34. Ghosh, S.K.; Pal, S. De-polymerization of LDPE plastic by *Penicillium simplicissimum* isolated from municipality garbage plastic and identified by ITSs locus of rDNA. *Vegetos* **2021**, *34*, 57–67. [CrossRef]
35. Daccò, C.; Girometta, C.; Asemoloye, M.D.; Carpani, G.; Picco, A.M.; Tosi, S. Key fungal degradation patterns, enzymes and their applications for the removal of aliphatic hydrocarbons in polluted soils: A review. *Int. Biodeterior. Biodegrad.* **2020**, *147*, 104866. [CrossRef]

36. Brown, N.A.; Ries, L.N.; Reis, T.F.; Rajendran, R.; Corrêa dos Santos, R.A.; Ramage, G.; Riaño-Pachón, D.M.; Goldman, G.H. RNAseq reveals hydrophobins that are involved in the adaptation of *Aspergillus nidulans* to lignocellulose. *Biotechnol. Biofuels* **2016**, *9*, 145. [CrossRef]
37. Kang, B.R.; Kim, S.B.; Song, H.A.; Lee, T.K. Accelerating the biodegradation of high-density polyethylene (HDPE) using *Bjerkandera adusta* TBB-03 and lignocellulose substrates. *Microorganisms* **2019**, *7*, 304. [CrossRef]
38. Spina, F.; Tummino, M.L.; Poli, A.; Prigione, V.; Ilieva, V.; Cocconcelli, P.; Puglisi, E.; Bracco, P.; Zanetti, M.; Varese, G.C. Low density polyethylene degradation by filamentous fungi. *Environ. Pollut.* **2021**, *274*, 116548. [CrossRef]
39. Hock, O.G.; De Qin, D.; Lum, H.W.; Hee, C.W.; Shing, W.L. Evaluation of the plastic degradation ability of edible mushroom species based on their growth and manganese peroxidase activity. *Curr. Top. Toxicol.* **2020**, *16*, 65–72.
40. Iiyoshi, Y.; Tsutsumi, Y.; Nishida, T. Polyethylene degradation by lignin-degrading fungi and manganese peroxidase. *J. Wood Sci.* **1998**, *44*, 222–229. [CrossRef]
41. Milstein, O.; Gersonde, R.; Huttermann, A.; Chen, M.J.; Meister, J.J. Fungal biodegradation of lignopolystyrene graft copolymers. *Appl. Environ. Microbiol.* **1992**, *58*, 3225–3232. [CrossRef]
42. Müller, R.J.; Schrader, H.C.G.; Profe, J.; Dresler, K.; Deckwer, W.D. Enzymatic degradation of poly(ethylene terephthalate): Rapid hydrolyse using a hydrolase from *T. fusca*. *Macromol. Rapid Commun.* **2005**, *26*, 1400–1405. [CrossRef]
43. Arkatkar, A.; Arutchelvi, J.; Sudhakar, M.; Bhaduri, S.; Uppara, P.V.; Doble, M. Approaches to enhance the biodegradation of polyolefins. *Open Environ. Eng. J.* **2009**, *2*, 68–80. [CrossRef]
44. Sheik, S.; Chandrashekar, K.; Swaroop, K.; Somashekarappa, H. Biodegradation of gamma irradiated low density polyethylene and polypropylene by endophytic fungi. *Int. Biodeterior. Biodegrad.* **2015**, *105*, 21–29. [CrossRef]
45. Ghosh, S.; Qureshi, A.; Purohit, H.J. Microbial degradation of plastics: Biofilms and degradation pathways Contam. In *Contaminants in Agriculture and Environment: Health Risks and Remediation*; Agro Environ Media: Haridwar, India, 2019; Chapter 14; pp. 184–199.
46. Yang, Y.; Yang, J.; Wu, W.M.; Zhao, J.; Song, Y.; Gao, L.; Yang, R.; Jiang, L. Biodegradation and mineralization of polystyrene by plastic-eating mealworms: Part 1. Chemical and physical characterization and isotopic tests. *Environ. Sci. Technol.* **2015**, *49*, 12080–12086. [CrossRef]
47. Yang, S.S.; Ding, M.Q.; Ren, X.R.; Zhang, Z.R.; Li, M.X.; Zhang, L.L.; Pang, J.W.; Chen, C.X.; Zhao, L.; Xing, D.F.; et al. Impacts of physical-chemical property of polyethylene on depolymerization and biodegradation in yellow and dark mealworms with high purity microplastics. *Sci. Total Environ.* **2022**, *828*, 154458. [CrossRef]
48. Hou, L.; Majumder, E.L.W. Potential for and distribution of enzymatic biodegradation of polystyrene by environmental microorganisms. *Materials* **2021**, *14*, 503. [CrossRef]
49. Woo, S.; Song, I.; Cha, H.J. Fast and facile biodegradation of polystyrene by the gut microbial flora of *Plesiophthalmus davidis* larvae. *Appl. Environ. Microb.* **2020**, *86*, e01361-20. [CrossRef]
50. Mahalakshmi, V.; Siddiq, A.; Andrew, S.N. Analysis of polyethylene degrading potentials of microorganisms isolated from compost soil. *Int. J. Pharm. Biol. Arch.* **2012**, *3*, 1190–1196.
51. Das, M.P.; Kumar, S. An approach to low-density polyethylene biodegradation by *Bacillus amyloliquefaciens*. *3 Biotech* **2015**, *5*, 81–86. [CrossRef]
52. Ojha, N.; Pradhan, N.; Singh, S.; Barla, A.; Shrivastava, A.; Khatua, P.; Rai, V.; Bose, S. Evaluation of HDPE and LDPE degradation by fungus, implemented by statistical optimization. *Sci. Rep.* **2017**, *7*, 39515. [CrossRef]
53. Jung, M.R.; Horgen, F.D.; Orski, S.V.; Rodriguez, C.V.; Beers, K.L.; Balazs, G.H.; Jones, T.T.; Work, T.M.; Brignac, K.C.; Royer, S.J.; et al. Validation of ATR FT-IR to identify polymers of plastic marine debris, including those ingested by marine organisms. *Mar. Pollut. Bull.* **2018**, *127*, 704–716. [CrossRef] [PubMed]
54. Álvarez-Barragán, J.; Domínguez-Malfavón, L.; Vargas-Suárez, M.; González-Hernández, R.; Aguilar-Osorio, G.; Loza-Tavera, H. Biodegradative activities of selected environmental fungi on a polyester polyurethane varnish and polyether polyurethane foams. *Appl. Environ. Microbiol.* **2016**, *82*, 5225–5235. [CrossRef] [PubMed]
55. Gómez-Méndez, L.D.; Moreno-Bayona, D.A.; Poutou-Piñales, R.A.; Salcedo-Reyes, J.C.; Pedroza-Rodríguez, A.M.; Vargas, A.; Bogoya, J.M. Biodeterioration of plasma pretreated LDPE sheets by *Pleurotus ostreatus*. *PLoS ONE* **2018**, *13*, e0203786. [CrossRef]
56. Yanto, D.H.Y.; Krishanti, N.P.R.A.; Ardiati, F.C.; Anita, S.H.; Nugraha, I.K.; Sari, F.P.; Laksana, R.P.B.; Sapardi, S.; Watanabe, T. Biodegradation of styrofoam waste by ligninolytic fungi and bacteria. *IOP Conf. Ser. Earth Environ. Sci.* **2019**, *308*, 012001. [CrossRef]
57. Gardes, M.; Bruns, T.D. ITS primers with enhanced specificity for basidiomycetes–application to the identification of mycorrhizae and rusts. *Mol. Ecol.* **1993**, *2*, 113–118. [CrossRef]
58. Altschul, S.F.; Gish, W.; Miller, W.; Myers, E.W.; Lipman, D.J. Basic local alignment search tool. *J. Mol. Biol.* **1990**, *215*, 403–410. [CrossRef]

59. Kalra, K.; Chauhan, R.; Shavez, M.; Sachdeva, S. Isolation of laccase producing *Trichoderma* spp. and effect of pH and temperature on its activity. *Int. J. Chemtech Res.* **2013**, *5*, 2229–2235.
60. Hosoya, T. Turnip peroxidase IV. The effect of pH and temperature upon the rate of reaction. *J. Biochem.* **1960**, *48*, 178–189. [CrossRef]
61. Casciello, C.; Tonin, F.; Berini, F.; Fasoli, E.; Marinelli, F.; Pollegioni, L.; Rosini, E. A valuable peroxidase activity from the novel species *Nonomuraea gerenzanensis* growing on alkali lignin. *Biotechnol. Rep.* **2017**, *13*, 49–57. [CrossRef]
62. Archibald, F.S. A new assay for lignin-type peroxidases employing the dye azure B. *Appl. Environ. Microbiol.* **1992**, *58*, 3110–3116. [CrossRef]
63. Stojković, D.; Gašić, U.; Uba, A.I.; Zengin, G.; Rajaković, M.; Stevanović, M.; Drakulić, D. Chemical profiling of *Anthriscus cerefolium* (L.) Hoffm., biological potential of the herbal extract, molecular modeling and KEGG pathway analysis. *Fitoterapia* **2024**, *177*, 106115. [CrossRef] [PubMed]
64. Popović, N.; Matekalo, D.; Stojković, D.; Skorić, M.; Gašić, U.; Božunović, J.; Milutinović, M.; Petrović, L.; Nestorović Živković, J.; Dmitrović, S.; et al. Transient expression of PRISEs and Trichoderma-mediated elicitation promote iridoid production in *Nepeta sibirica* L. *Plant Physiol. Biochem.* **2025**, *225*, 109986.
65. Demšar, J.; Curk, T.; Erjavec, A.; Gorup, Č.; Hočevar, T.; Milutinović, M.; Možina, M.; Polajnar, M.; Toplak, M.; Starič, A.; et al. Orange: Data mining toolbox in python. *J. Mach. Learn. Res.* **2013**, *14*, 2349–2353.
66. Olmos, D.; Martín, E.V.; González-Benito, J. New molecular-scale information on polystyrene dynamics in PS and PS–$BaTiO_3$ composites from FTIR spectroscopy. *Phys. Chem. Chem. Phys.* **2014**, *16*, 24339–24349. [CrossRef]
67. Meenashi, S.; Umamaheswari, S. FTIR analysis of bacterial mediated chemical changes in Polystyrene foam. *Ann. Biol. Res.* **2016**, *7*, 55–61.
68. Bhutto, A.; Vesely, D.; Gabrys, B. Miscibility and interactions in polystyrene and sodium sulfonated polystyrene with poly(vinyl methyl ether) PVME blends. Part II. FTIR. *Polymer* **2003**, *44*, 6627–6631. [CrossRef]
69. Ilijin, L.; Nikolić, M.V.; Vasiljević, Z.Z.; Todorović, D.; Mrdaković, M.; Vlahović, M.; Matić, D.; Tadić, N.B.; Perić-Mataruga, V. Sourcing chitin from exoskeleton of *Tenebrio molitor* fed with polystyrene or plastic kitchen wrap. *Int. J. Biol. Macromol.* **2024**, *268*, 131731. [CrossRef]
70. Peng, B.Y.; Su, Y.; Chen, Z.; Chen, J.; Zhou, X.; Benbow, M.E.; Zhang, Y. Biodegradation of polystyrene by dark (*Tenebrio obscurus*) and yellow (*Tenebrio molitor*) mealworms (Coleoptera: Tenebrionidae). *Environ. Sci. Technol.* **2019**, *53*, 5256–5265. [CrossRef]
71. Jiang, S.; Su, T.; Zhao, J.; Wang, Z. Biodegradation of polystyrene by *Tenebrio molitor*, *Galleria mellonella*, and *Zophobas atratus* larvae and comparison of their degradation effects. *Polymers* **2021**, *13*, 3539. [CrossRef]
72. Uzan, E.; Nousiainen, P.; Balland, V.; Sipila, J.; Piumi, F.; Navarro, D.; Asther, M.; Record, E.; Lomascolo, A. High redox potential laccases from the ligninolytic fungi *Pycnoporus coccineus* and *Pycnoporus sanguineus* suitable for white biotechnology: From gene cloning to enzyme characterization and applications. *J. Appl. Microbiol.* **2010**, *108*, 2199–2213.
73. Cañas, A.I.; Camarero, S. Laccases and their natural mediators: Biotechnological tools for sustainable eco-friendly processes. *Biotechnol. Adv.* **2010**, *28*, 694–705. [CrossRef] [PubMed]
74. Anastasi, A.; Tigini, V.; Varese, G.C. The bioremediation potential of different ecophysiological groups of Fungi. In *Fungi as Bioremediators. Soil Biology*; Goltapeh, E., Danesh, Y., Varma, A., Eds.; Springer: Berlin/Heidelberg, Germany, 2013; Volume 32, pp. 29–49.
75. Piontek, K.; Smith, A.T.; Blodig, W. Lignin peroxidase structure and function. *Biochem. Soc. Trans.* **2001**, *29*, 111–116. [CrossRef] [PubMed]
76. Hofrichter, M. Review: Lignin conversion by manganese peroxidase (MnP). *Enzym. Microb. Technol.* **2002**, *30*, 454–466. [CrossRef]
77. Albertsson, A.C.; Andersson, S.O.; Karlsson, S. The mechanism of biodegradation of polyethylene. *Polym. Degrad. Stab.* **1987**, *18*, 73–87. [CrossRef]
78. Khatoon, N.; Jamal, A.; Ali, M.I. Lignin peroxidase isoenzyme: A novel approach to biodegrade the toxic synthetic polymer waste. *Environ. Technol.* **2019**, *40*, 1366–1375. [CrossRef]
79. Wong, D.W.S. Structure and action mechanism of ligninolytic enzymes. *Appl. Biochem. Biotechnol.* **2009**, *157*, 174–209. [CrossRef]
80. Faison, B.D.; Kirk, T.K.; Farrell, R.L. Role of veratryl alcohol in regulating ligninase activity in *Phanerochaete chrysosporium*. *Appl. Environ. Microbiol.* **1986**, *52*, 251–254. [CrossRef]
81. Caramelo, L.; Martínez, M.J.; Martínez, A.T. A search for ligninolytic peroxidases in the fungus *Pleurotus eryngii* involving α-keto-γ-thiomethylbutyric acid and lignin model dimers. *Appl. Environ. Microbiol.* **1999**, *65*, 916–922. [CrossRef]
82. Kim, W.; Lee, B.; Park, J.; Kim, H.J.; Cheong, H. Comparative antioxidant activity and structural feature of protocatechuic acid and phenolic acid derivatives by DPPH and intracellular ROS. *Lett. Drug Des. Discov.* **2018**, *15*, 612–620. [CrossRef]
83. Rice-Evans, C.A.; Miller, N.J.; Paganga, G. Structure-antioxidant activity relationships of flavonoids and phenolic acids. *Free Radic. Boil. Med.* **1996**, *20*, 933–956. [CrossRef]

84. Yang, S.S.; Wu, W.M.; Brandon, A.M.; Fan, H.Q.; Receveur, J.P.; Li, Y.; Wang, Z.Y.; Fan, R.; McClellan, R.L.; Gao, S.H.; et al. Ubiquity of polystyrene digestion and biodegradation within yellow mealworms, larvae of *Tenebrio molitor* Linnaeus (Coleoptera: Tenebrionidae). *Chemosphere* **2018**, *212*, 262–271. [CrossRef] [PubMed]
85. Yu, X.; Zhang, Y.; Chen, S.; Chen, S.; Wan, C.; Wang, Y.; Zou, L.; Peng, L.; Ye, L.; Li, Q. Study on the degradation efficiency and mechanism of polystyrene microplastics by five kinds of edible fungi. *J. Hazard. Mater.* **2025**, *492*, 138165. [CrossRef]

polymers

MDPI

Article

Preparation of Polystyrene Nanoparticles with Environmental Relevance Using a Gradual Degradation Method

Hisayuki Nakatani [1,2,*], Mika Asano [1], Masaki Sakamoto [1], Suguru Motokucho [1,2], Anh Thi Ngoc Dao [1], Hee-Jin Kim [3], Mitsuharu Yagi [3] and Yusaku Kyozuka [2]

[1] Chemistry and Materials Engineering Program, Graduate School of Integrated Science and Technology, Nagasaki University, 1-14 Bunkyo-machi, Nagasaki 852-8521, Japan; bb54124102@ms.nagasaki-u.ac.jp (M.A.); bb54125111@ms.nagasaki-u.ac.jp (M.S.); motoku@nagasaki-u.ac.jp (S.M.); anh.dao@nagasaki-u.ac.jp (A.T.N.D.)

[2] Organization for Marine Science and Technology, Nagasaki University, 1-14 Bunkyo-machi, Nagasaki 852-8521, Japan; kyozuka@nagasaki-u.ac.jp

[3] Fisheries Bioresources Program, Graduate School of Integrated Science and Technology, Nagasaki University, 1-14 Bunkyo-machi, Nagasaki 852-8521, Japan; heejin@nagasaki-u.ac.jp (H.-J.K.); yagi-m@nagasaki-u.ac.jp (M.Y.)

* Correspondence: h-nakatani@nagasaki-u.ac.jp

Abstract: This study investigates the environmental degradation of polystyrene (PS) microparticles and flakes using a gradual degradation method. The concentration of $SO_4\bullet^-$ decreased exponentially, simulating the environmental conditions. The nanofragment size of PS particles evolved dynamically, fluctuating from below 250 nm at 3 days to 300–500 nm at 6 days, then forming two peaks below 200 nm at 9 days, before shifting to a single peak below 100 nm at 12 days. At 15 days, the distribution expanded to two peaks between 500 nm and 200 nm. The polydispersity index (PDI) varied unpredictably, and fragments below 100 nm fluctuated between 10 and 50 nm independent of time. SEM analysis revealed an initial peeling process, with the outermost layer peeling off. The core size of the PS particles decreased rapidly from 11,000 nm to 2500 nm within 6 days and stabilized at 1000 nm after 9 days. The PS flakes showed minimal shape change until 24 days, but surface roughness increased by 30 days, leading to fragmentation. By 42 days, the flakes partially broke into ca. 100 μm pieces. The initial morphology significantly influenced the breakdown pattern, suggesting multiple breakdown mechanisms other than peeling.

Keywords: microplastics; mechanism; sea; polystyrene; degradation method; environmental relevance

1. Introduction

The world's oceans are polluted with plastic [1–5]. When plastic products are thrown into the environment as waste, they degrade under the sun's rays and due to mechanical stress from wind and waves, breaking down into microplastics (MP) smaller than 5 mm [6,7] and then further into nanoplastics (NP) smaller than 1000 nm [8,9]. The recovery and confirmation of NPs in the environment require extremely precise handling [10]. Nevertheless, some studies have confirmed the presence of NP particles in the oceans [8,9], raising concerns due to their unprecedented stability and unknown effects on human health. Research on NPs has primarily focused on polystyrene (PS) particles and mechanically ground commodity plastics [11–13]. One study has examined the ecological impact of NPs, which can interfere with lung surfactant function [12]. PS-NPs have shown reproductive toxicity in animal models, which is linked to oxidative stress and mitochondrial

dysfunction, potentially affecting human reproduction [13]. NPs are found in various environmental matrices, including air, water, soil, and food, and they can enter the human body through inhalation, ingestion, or dermal absorption [12,14]. They may reach the brain, potentially causing diseases by initiating molecular or cellular reactions, damaging the blood–brain barrier, inducing oxidative stress and inflammatory responses, affecting acetylcholinesterase activity, causing mitochondrial dysfunction, and interfering with autophagy [15]. These findings highlight the significant health risks posed by NPs and underscore the need for further research and awareness [12–15].

To clearly demonstrate the actual impact of NPs on living organisms, including humans, it is essential to establish a method for creating NP models with environmental relevance [9]. Actual NPs are mainly composed of fragments of plastic waste degraded by exposure to sunlight and depend on the amount of industrial plastic produced. Therefore, it is necessary to generate NP models by fragmenting polypropylene (PP), polyethylene (PE), and polystyrene (PS). Considering the environmentally related mechanistic aspects of degradation, it is necessary to use autoxidation degradation [16,17]. A method using UV light or a xenon lamp has often been used as a method of micronization by autoxidation [18,19]. Certainly, since it is sunlight that initiates autoxidation in nature, the use of light to induce autoxidation is environmentally relevant. On the other hand, conventional photo-induced autoxidation is continuous, where light is irradiated to cause degradation at a constant rate. This is because degradation by light is slow, and acceleration is required to achieve the desired effect. The constant exposure to light is very different from natural degradation. The sun sets, and the intensity of sunlight varies with the seasons. Sunlight is also blocked by biofilm and dust. A particularly important consideration is ocean degradation. Plastic debris slowly sinks into the ocean, and the intensity of degradation by sunlight gradually decreases. There should be a difference in the fragmentation of plastics between such "uniform degradation" and "gradual degradation", but no one has yet investigated what kind of fragmentation behavior occurs in plastics subjected to "gradual degradation". Controlling autoxidation degradation is difficult, but it is possible if the stability of the hydroperoxide group, which is the rate-limiting step, and the generation of active oxygen, which is the initiating species, can be controlled [20,21]. In our previous study [22], we conducted PP degradation in seawater using an advanced oxidation process (AOP) with sulfate ion radical ($SO_4\bullet^-$) as a highly efficient initiator for plastic degradation. The combination of seawater and the $SO_4\bullet^-$ initiator resulted in the excellent acceleration of the degradation process under pH control. This accelerated degradation method can be applied not only to PP, but also to PS, which has the same autoxidation degradation mechanism [23]. Therefore, we believe that it is possible to reproduce the "gradual degradation" with some degree of acceleration.

In this study, the gradual degradation behavior of PS microparticles under environmental conditions was investigated while preventing aggregation during the degradation using Triton® X-114 surfactant. Particle size fluctuations were measured at various intervals, and complex changes in the size distribution of PS nanofragments were studied. The breaking down mechanism of PS microparticles was estimated by analyzing scanning electron microscope (SEM) images. In addition, by using PS flake samples as a model for marine plastic waste, gradual degradation was performed, and the degradation process was studied. The effect of sample shape on fragmentation was assessed by comparing the behavior of PS microparticles and flakes in detail. This study is the first to demonstrate its application to PS in a simulated marine environment with a focus on environmentally relevant degradation behavior. Our experimental design introduces a novel approach to simulate real-world environmental conditions by controlling the concentration of $SO_4\bullet^-$ to

mimic the attenuation of sunlight due to natural factors such as cloud cover, diurnal cycles, and particle sinking.

2. Materials and Methods

2.1. Materials

PS particles (Standard Particle PHARM-TROLTM Series 15 μm) were obtained from Thermo Fisher Scientific K.K. (Tokyo, Japan) with a size of ca. 15 μm, and their particle count was 3800 ± 15%/mL. Wako Pure Chemical Industries (Osaka, Japan) provided potassium persulfate ($K_2S_2O_8$), PS pellets (the weight-average molecular weight and molecular weight distribution were 3.5×10^5 and 2.1, respectively.), and Triton(R) X-114 (Polyethylene glycol tert-octylphenyl ether). The seawater used was artificial (Gex artificial saltwater), which was purchased form Amazon.co.jp (Tokyo, Japan. https://www.amazon.co.jp/-/en/GEX-Gex-Seawater-6-5-Pieces/dp/B09R9X4Y43?th=1, accessed on 5 September 2024).

2.2. Pulverization and Film Molding Methods Using PS Pellets

The PS pellet pulverization was carried out with a freeze crusher (TPH-01, AS ONE Co., Osaka, Japan) with liquid nitrogen to obtain the powder. The powder was passed through a sieve with 100 mesh (Filtration particle size: ca. 250 μm) and was employed as a pulverized sample. The specimen for PS film was ca. 30 mm × 30 mm × 0.075 mm. The film was obtained by compression molding at 180 °C under 10 MPa for 11 min.

2.3. Gradual Degradation Using Sulfate Ion Radicals in Seawater

The gradual degradation of the PS particles was carried out by means of an advanced oxidation process (AOP) with sulfate ion radicals in the artificial seawater. The procedure was in accordance with our previous reports [22,23]. (1) The 25 mL PS particle solution (containing a total of approximately 95,000 ± 15% particles) was put into a 100 mL glass vessel containing 20 mL of seawater solution with 0.54 g $K_2S_2O_8$ at ca 65 °C for 12 h under stirring with a stirrer tip speed of ca 100 rpm. (2) To compensate for the consumption of the oxidizing agent, 20 mL of the same $K_2S_2O_8$ seawater solution was added, and the degradation was carried out for 12 h under the same conditions. (3) As the 20 mL $K_2S_2O_8$ seawater solution was added every 12 h, the volume of the solution continued to increase; thus, the entire solution was transferred to a larger vessel as appropriate. The degradations started again under the same conditions. All gradual degradation was carried out at the same temperature (65 °C) and with a stirring speed of 100 rpm. The pH value of the solution was changed from 8.2 to 3 during each set (the pH of the seawater was initially 8.2, and the $SO_4\bullet^-$ was gradually converted to SO_4^{2-}, reducing the pH of the seawater solution to 3 at the time of the daily exchange [22,23]). The samples that had reached the predetermined degradation time were filtered using filter paper with a pore size of 200 nm, rinsed with a small amount of pure water, and then dried in a vacuum dryer at 60 °C for 7 h before being used for each measurement.

The gradual degradation of the 10 mg pulverized PS pellet was first put into a 100 mL glass vessel containing 20 mL of seawater solution with 0.54 g $K_2S_2O_8$ at ca 65 °C for 12 h under stirring with a stirrer tip speed of ca 100 rpm, and then the same procedure as in the gradual degradation method of PS particles was used. The filtration and drying methods were also the same as those used for the PS particles.

2.4. Characterization and Analysis

Scanning electron microscopy (SEM) analysis was conducted using a JSM-7500FAM (JEOL, Tokyo, Japan) at an electron beam voltage of 5.0 kV. The working distance was

approximately 3 × 4 mm. The samples were placed in a drying oven maintained at 27 °C for 30 min and sputter-coated with gold before undergoing SEM imaging.

The hydrodynamic size was determined under pure water suspension at 20 °C using an ELSZ-2000ZS dynamic light scattering (DLS) analyzer manufactured by Otsuka Electronics (Osaka, Japan). The hydrodynamic size of the fragment contained in the conical tube after processing was determined by measuring the path length and width of a borosilicate glass standard fluorescence cell (10 mm each) and the capacity (3.5 mL). The DLS analyzer was used to measure the particle size distribution. The range of sizes was measured from 0.6 nm to ca. 15 μm, with the DLS analyzer operating at a temperature of 20 °C and using pure water (viscosity 0.8878 cP, refractive index 1.3328) as the solvent for each measurement. The DLS analyzer performed 25 accumulations for each sample.

The transform infrared spectra of 16 scans were measured with a Fourier transform infrared spectrometer Jasco FT-IR 660 plus (Jasco, Tokyo, Japan) with a resolution of 4 cm^{-1} over the entire mid-IR range (400–4000 cm^{-1}).

2.5. Statistical Processing

Hydrodynamic size distribution and polydispersity index (PDI) were calculated using a software (ELSZ-2000 Version 7.16, Otsuka Electronics Co., Ltd., Osaka, Japan) supplied with the DLS.

3. Results and Discussion

3.1. Peeling and Breaking Down Behavior of PS Particles by Gradual Degradation

As shown in Figure 1, the concentration of $K_2S_2O_8$, i.e., $SO_4\bullet^-$ decreased exponentially with the increasing degradation time. By controlling the concentration of $SO_4\bullet^-$ in this way, we were able to successfully observe the gradual degradation behavior of PS particles by reducing the degradation rate as the degradation time increased. This gradual degradation behavior simulated environmental degradation due to the attenuation of sunlight caused by factors such as changing day/night, sunlight being blocked by clouds, and the sinking of irradiated targets to the seafloor. Figure 2 shows the hydrodynamic size distribution of all PS particles at each degradation time. The hydrodynamic size of degraded PS particles shows complex distribution behavior at each degradation time. After 3 days of degradation, there were particles close to the original particle size of 8 to 15 μm, but after the 6 days, they decreased to 3 μm. After 6 days of degradation, relatively large particles of a few microns did not show a clear size dependence on degradation time, but instead they showed repeated size increases and decreases. The peak on the left of the graph for each degradation time indicates particles of several hundred nanometers, which have a complex size distribution. For degradation times of 3 and 9 days, the corresponding peaks appear as double peaks, while for other degradation times, the peaks appear as single peaks. When measuring DLS, 1% of the surfactant [Triton(R) X-114] was added. In the previous research [23], the same concentration of Triton(R) X-114 surfactant was successful in suppressing the formation of aggregation structures in low-density polyethylene (PE) nanoparticle fragments produced by similar AOP degradation. According to the report by Li et al. [24], the most important factor for the nano-dispersibility of PS nanoparticles is the charge on the surface of the nanoparticles. The charge on the surface of the nanoparticles is negative and is derived from oxidation compounds containing carbonyl groups, etc., which are produced by decomposition. The longer the degradation time, the greater the absolute value of the negative charge is expected to be. Increasing the absolute value of the negative charge on the surface increases the repulsive force between nanoparticles, making them less likely to aggregate. Therefore, the use of surfactants is essential in cases where the degree of degradation is low. In this study, the use of Triton(R) X-114 surfactant is essential

to suppress the aggregation of PS fragments because the ability to degrade PS decreases over time. As shown in Figure 2, the number of PS particles in the hundreds of nanometer size range continues to increase and decrease as the degradation time progresses. For example, at a degradation time of 3 days, the particle size distribution showed two splits at 250 nm or less for the relevant particle size, but at a degradation time of 6 days, it showed a slightly broad monodisperse particle size distribution from 300 nm to 500 nm. After the degradation time of 9 days, the particle size distribution showed two peaks of less than 200 nm, and after 12 days it further decreased to a single peak of less than 100 nm. After 15 days, it increased again to a two-peak distribution from 500 nm to 200 nm.

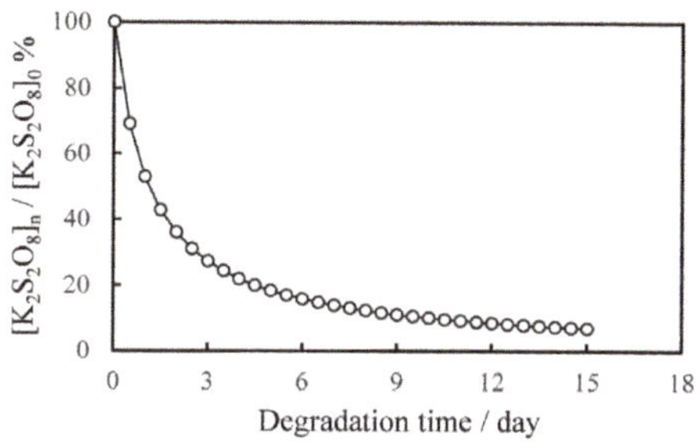

Figure 1. Dependence of degradation time on $K_2S_2O_8$ concentration ratio: $[K_2S_2O_8]_0$ = 44.4×10^{-3} mmol/L.

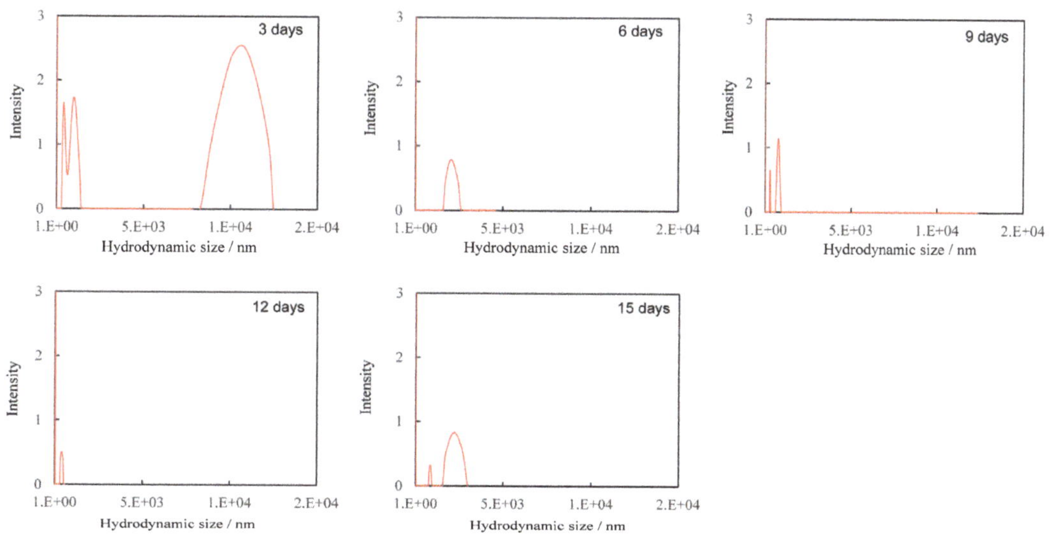

Figure 2. Hydrodynamic size distribution of all PS particles at each degradation time.

The dependence of PDI on degradation time is summarized in Table 1. The PDI value repeatedly increased and decreased with degradation time, but no clear dependence could be observed. Such complex changes in particle size distribution over time suggest that new fine fragments of different sizes are being created over time, or that further refinement is occurring, resulting in fragments that are too small to measure. Figure 3 shows the hydrodynamic size distribution of fragments smaller than 100 nm at each degradation time. Even in the smaller nanoscale region, the size of the PS particle fragments did not show any dependence on degradation time, as was the case for the larger size, and the width of the size distribution repeatedly increased and decreased from 10 nm to 50 nm. These changes in

size distribution and PDI are independent of degradation time and are therefore thought to be related to the disappearance of fragments that reached the measurement limit due to the generation of new fine fragments by further fragmentation. In fact, the change in particle fragmentation size distribution is not related to degradation time, suggesting that the degree of degradation in the fragments does not depend on the degradation time. It is necessary to accept the behavior that when degradation progresses to a certain degree, the surface layer peels off and a new, less degraded surface appears in order to explain the change in particle size distribution. Some research groups have reported that the photodegradation of polyolefins causes the degrading layer to detach from the surface [25,26]. Figure 4 shows the SEM image of PS particles degraded for 1 day, schematically illustrating the peeling from the surface layer. It can be observed that it is starting to peel off. The gradual degradation causes the core and fragment parts. The SEM photograph captures the initial stage of the peeling process, showing that the outermost layer appears to be peeling away. The diameter of PS particles immediately after peeling was 15.34 μm, which was almost the same as the diameter before the gradual degradation. No decrease in micro size was observed in the PS particle before and after the gradual degradation, i.e., before and after the peeling away. The fact that the size distribution and PDI of fragments do not depend on degradation time can be explained by the decrease in degradation rate due to radical initiators, which decrease exponentially with increasing degradation time. In fact, as shown in Figure 5, when the fragments are divided into a fragment part (1000 nm less) and a core part (1000 nm or more), the plot of the average fragment size (average hydrodynamic size) of the fragment part versus degradation time shows an exponential decreasing trend versus degradation time. After the degradation time of 3 days, the average hydrodynamic size of the fragment part was ca. 200 nm, and after 9 days, it decreased to ca. 20 nm. However, it remained almost constant during the subsequent degradation time. This is because the peeling part gradually degrades and becomes finer in accordance with the decrease in $SO_4{\bullet}^-$ concentration and because the newly peeling part from the core becomes smaller in response to the decrease in concentration. Interestingly, the average hydrodynamic size of the core part also decreases exponentially with degradation time. The mechanism by which nanometer-thick fragments peel off from the approximately 15 μm core part as degradation progresses cannot explain such an exponential reduction in hydrodynamic size. This is because, as the radical concentration decreases, the rate of degradation slows, making it more difficult for the material to peel off, and the fragments that do peel off become smaller and thinner. The result that the hydrodynamic size of the core part shows an exponential decrease with respect to the degradation time suggests the existence of fragmentation mechanisms other than peeling. In photo-degradation reactions, it is known that the progression of degradation slows down toward the interior due to the diffusion of oxygen [27,28]. White et al. reported that, in ultraviolet-degraded polypropylene (PP), cracks in the surface layer that has undergone degradation are arrested on reaching the ductile material in the interior, which has not been degraded or has been degraded to a low degree and retains its toughness [29]. In our degradation system using $SO_4{\bullet}^-$ in seawater, oxygen permeability is similar, with the oxygen concentration decreasing from the surface to the interior of PS particles. Depth degradation is strongly influenced by the diffusion behavior of oxygen, even in degradation caused by this system. PS is an amorphous polymer, so it does not undergo chemical crystallization when it degrades and, therefore, is not affected by residual stress in the direction of cracking. The depth and density of cracks appear to be relatively uniform across the surface of the sphere. The peeling occurs in two dimensions, and the thickness of the peeling area is considerably smaller than its length in both the vertical and horizontal directions (see Figure 4). After peeling, cracks in the depth direction remain on the surface. As the $SO_4{\bullet}^-$ concentration decreases, the

next peeling area becomes smaller, but as the radicals penetrate from the cracks into the depth direction, the depth of the cracks gradually increases. Obviously, the penetration of radicals into the crack causes degradation not only in the depth direction, but also in the vicinity, leading to the embrittlement of the entire matrix. When the matrix exceeds a certain threshold of brittleness due to shear caused by the stirring of seawater and collisions between particles, it breaks down. As shown in Figure 5, the average hydrodynamic size of the core decreased rapidly from ca. 11,000 nm at the 3-day degradation time to ca. 2500 nm at the 6-day degradation time. This behavior can be explained by considering the breakdown mechanism. Figure 6 shows the SEM image of PS particles degraded for 3 days, schematically illustrating the breaking down. In the SEM image, the PS particle appears as an irregular sphere with defects here and there, with a diameter of ca. 5 μm. This sphere is the core part that broke into large fragments during the breakdown, which is evidence of the existence of a breakdown mechanism. As shown in Figure S1, the diameter of the core decreased from 10 μm to 1 μm as the degradation time increased up to 9 days and remained at approximately 1 μm for longer degradation times. On the other hand, the shape remained spherical, albeit irregular, regardless of the degradation time. The result that the shape was maintained suggests that a repetitive mechanism occurred in which the interior, which had been degraded to a low degree, was exposed to the surface by the breakdown, followed by further peeling and breaking down. Plastic that has weathered under natural conditions, especially in coastal areas, is known to have a distinctive pitted texture on its surface [30–33]. Plastic degradation tends to produce pit textures in marine environments where UV radiation and mechanical erosion are minimal [30]. Cooper et al. inferred that the pits observed on the surface of PE debris collected along the coast are caused by collisions [31]. In addition, although no damage was observed in other plastics, the presence of small areas of severe oxidation (which Copper et al. referred to as "initial degradation sites" and "preferential degradation") was also noted [31]. Their information suggests that the PS particle breakdown phenomenon observed in the gradual degradation we found may complement the pit formation mechanism. Plastic weathering tests are generally performed in the air. A light source, such as a xenon lamp, is used to continuously irradiate the plastic at a constant intensity to cause degradation. Due to a number of factors, the degradation rate is expected to decrease gradually under natural conditions. In particular, the effects of seawater on coastal weathering cannot be ignored. Aqueous Cl^- functions as an inhibitor in the photooxidation of polymers such as polyolefin in seawater [34]. The pits observed on the surface of PE debris found on the coast are thought to have been generated by a mechanism similar to the breakdown observed in PS particles in this study. In areas of the PE debris surface that would be partially submerged in seawater, degradation due to exposure to sunlight progressed, resulting in peeling and cracking. The degradation repeatedly stopped and progressed, and cracks in the depth direction accumulated at specific locations where degradation could progress. Eventually, complex mechanical stresses such as shear forces from waves and friction with sand caused breakdowns at these specific locations, and the collapsed sections were washed away by waves, forming the pits of PE. However, whether or not a pit-like texture forms appears to depend on the type of polymer. For example, on the coast, deep and distinct crack textures are mainly observed in PP debris [31]. As described above, the PS particles disintegrated as if the outer skin had been peeled off. They retained their spherical shape. The shape of the texture is formed by degradation, and mechanical stimulation depends on the characteristics of the degraded polymer. The most important factor in determining the shape of a texture is whether it is crystalline or not. This is because crystalline polymers such as PE and PP undergo chemi-crystallization, which leads to embrittlement when they are degraded [27,28]. The depth density of cracks and the

brittleness of the matrix polymer increase as degradation progresses, causing the degraded areas to fragment under mechanical stress. In this case, whether the degraded part breaks into fine particles and is removed to form pits, as in PE, or is removed in relatively large rock-like pieces to form deep, distinct crack-like textures, as in PP, would be determined by the crystallization speed and glass transition temperature of each polymer. On the other hand, in amorphous polymers such as PS, chemi-crystallization does not occur; thus, cracks in the depth direction occur at relatively uniform depths, and as degradation progresses, the density of these cracks increases, and cracks develop horizontally at a certain depth and connect two-dimensionally. Therefore, when shear forces like waves are applied, spheres, which tend to degrade at a uniform rate due to their shape, peel off in blocks of uniform thickness, and their diameter decreases in a similar manner while maintaining their shape.

Table 1. Dependence of polydispersity index (PDI) on degradation time.

Deg. Time	3 Days	6 Days	9 Days	12 Days	15 Days
PDI	0.246 ± 0.065	0.163 ± 0.039	0.212 ± 0.026	0.176 ± 0.038	0.154 ± 0.019

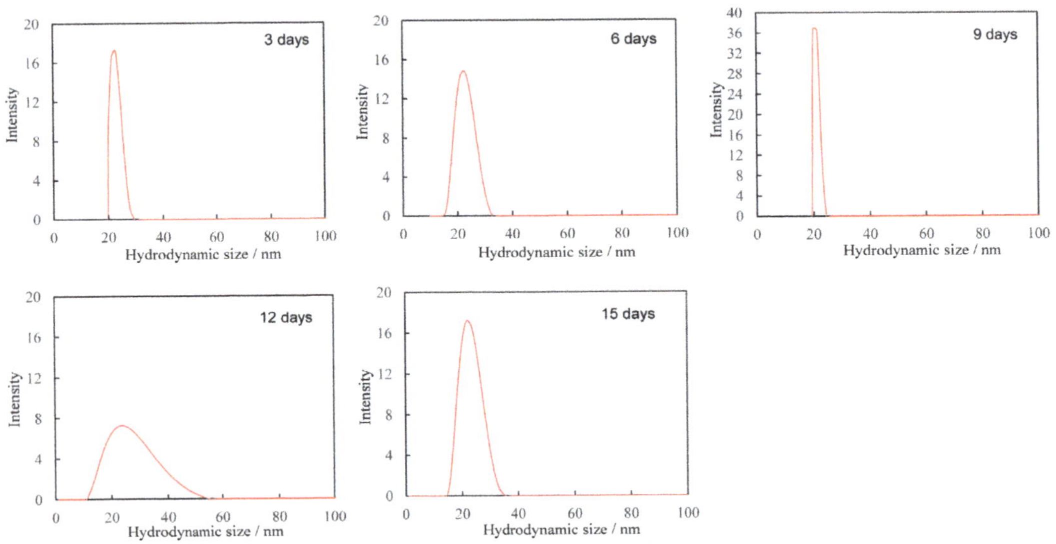

Figure 3. Hydrodynamic size distribution of particle fragments smaller than 100 nm at each degradation time.

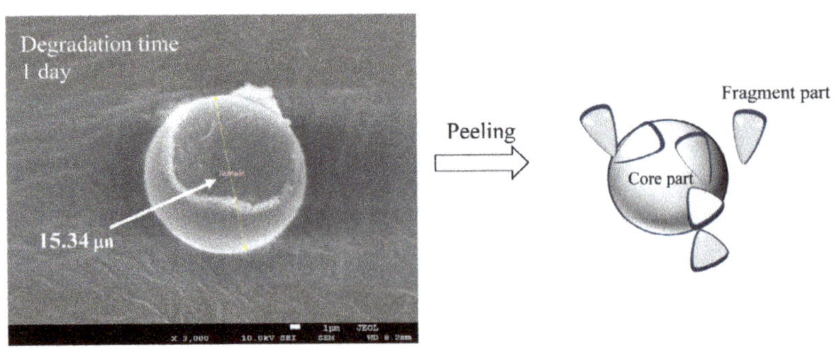

Figure 4. SEM image of a gradually degraded PS particle and image of peeling.

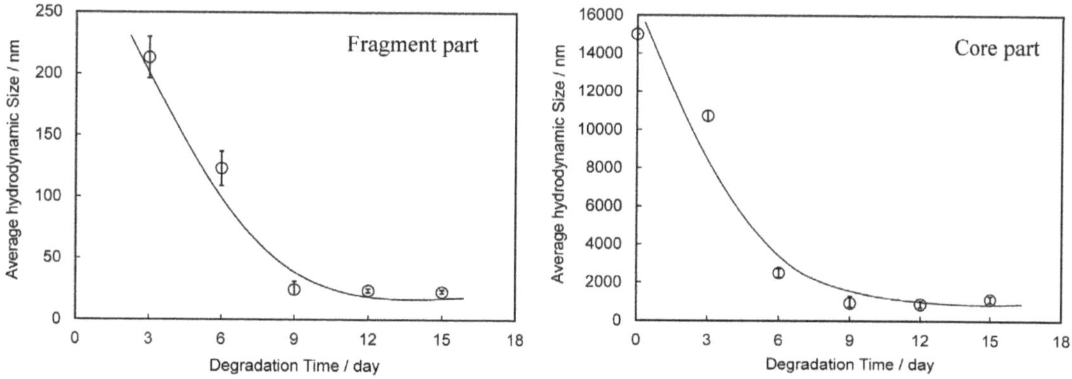

Figure 5. Degradation time dependences of average hydrodynamic size in fragment and core parts.

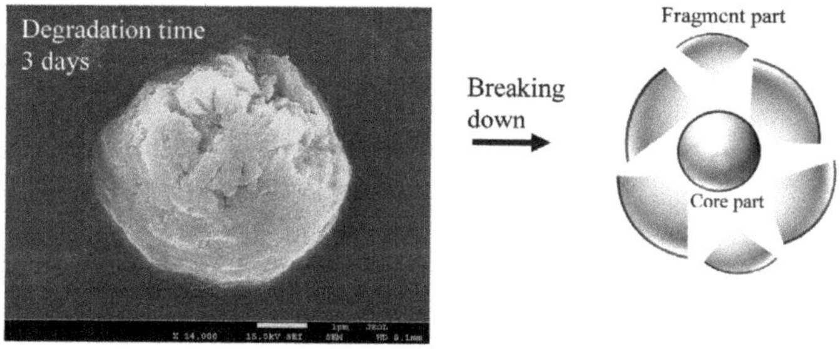

Figure 6. SEM image of a gradually degraded PS particle and image of breakdown.

3.2. Fragmentation Behavior of PS Pulverized Pellets by Gradual Degradation

From an environmental impact perspective, spherical plastic waste is referred to as primary microplastics, but flake-shaped secondary microplastics generated by degradation and/or mechanical stress are clearly present in much greater quantities in the natural environment. It is possible to take spectroscopic measurements using FT-IR for flake-shaped specimens, which enables confirmation of the degradation process. Therefore, for the plastic specimens used in this study, which simulates the gradual degradation under natural conditions, flake-shaped specimens are preferable to spherical specimens. Plastic debris released into the ocean becomes covered with microalgae and other organisms, causing what is known as "biofouling". As biofouling progresses, the surface of the debris becomes thickly covered with a biofilm, preventing sunlight from reaching the surface of the plastic debris [35]. Furthermore, as the overall weight increases, the plastic debris covered with biofouling gradually sinks into the sea. As the sunlight is gradually blocked and the plastic debris sink into the sea, the rate of degradation gradually decreases. This is exactly what is meant by "gradual degradation". Using PS pulverized pellet flakes (major axis: a few hundred micrometers) as a model for degradation progression in the sea, the gradual degradation was used to investigate changes in shape and size by SEM observation. Confirmation of the gradual degradation (autoxidation) reaction was performed using FT-IR measurement. In the PS pulverized pellet flakes that underwent gradual degradation for 21 days, peaks assigned to carbonyl groups and hydroperoxide groups were observed at around 1720 cm^{-1} and 3500 cm^{-1}, respectively, confirming that autoxidation had occurred (see Figure S2). As shown in Figure 7, no clear shape changes dependent on degradation time were observed until the 24th day of degradation. After 30 days of degradation, the flake surface became rough, and fragmentation could be observed peeling off the surface,

similar to the PS particles. However, compared to the PS particles, there were significant differences in the breakdown behavior. In the PS flakes, broken pieces ca. 100 μm in size separated by crevice-like cracks were observed in samples after 42 days of degradation time. In the case of PS particles, they retained their spherical shape after the breakdown and became smaller in a similar manner, but in the case of flakes, they did not become similar in shape but rather cracked discontinuously and split into smaller pieces. The PS particles were not oriented, but the flakes retained residual stress from pulverization and were oriented in a certain direction. Therefore, cracks tended to occur in the orientation direction. The PS flakes measuring a few hundred micrometers in size were broken down into pieces measuring ca. 100 μm after 42 days of decomposition. At this time, the $SO_4{\bullet}^-$ concentration had decreased to 2.6% of the original value, and the degradation intensity (speed) had significantly decreased. This indicates that the growth of crevices is not due to chemical factors but is rather strongly dependent on mechanical factors such as residual stress and shear stress from seawater. Pabortsava et al. reported that PE, PP, and PS particles with sizes ranging from 32 μm to 651 μm were suspended within the top 200 m of the Atlantic Ocean, with a total estimated quantity from 11.6 to 21.1 million tons [36]. Interestingly, the size of the fragments is similar to that of the PS pieces that broke off and separated. It is likely that suspended PE, PP, and PS particles in the sea, where sunlight does not easily reach, are generated by the same mechanical mechanism, resulting in similar sizes.

Figure 7. SEM photographs of gradually degraded PS pulverized pellets at various degradation times.

This method has great potential for use in environmental monitoring and toxicity research. In environmental monitoring, it can detect and quantify pollutants, assess ecosystem health, and track changes over time. In toxicity research, this method allows scientists to evaluate the effects of various substances on biological systems and provides insight into potential risks to humans and the environment. However, obtaining reliable and reproducible results requires considering limitations such as sensitivity and specificity and the need for standardized protocols. This requires conducting numerous preliminary experiments, which necessitates producing a large number of NP samples. Based on the mechanisms of NP formation identified in this experiment, we plan to develop methods for large-scale NP production in the future.

4. Conclusions

The concentration of $SO_4\bullet^-$ decreased exponentially with degradation time, simulating environmental conditions. The size of PS spherical particle distribution changed in a complex manner. The usage of Triton(R) X-114 surfactant effectively prevented the agglomeration of fragments generated by the degradation. As degradation progressed, the nanofragment sizes fluctuated in the following way: below 250 nm at 3 day of the degradation time, 300–500 nm at 6 days, two peaks under 200 nm at 9 days, and a single peak below 100 nm at 12 days. After 15 days, a two-peak distribution reappeared, ranging from 500 nm to 200 nm. The PDI value repeatedly increased and decreased with the degradation time, but no clear dependence was observed. The fragments smaller than 100 nm, with size distribution fluctuating between 10 nm and 50 nm, did not depend on degradation time. These results suggest that new nanofragments were generated as larger fragments disappeared. The degree of degradation did not depend on time alone but rather on the peeling of the surface layer, revealing a less degraded surface. The gradual degradation formed the core and fragment parts. The SEM photograph captured the initial stage of the peeling process, showing that the outermost layer was peeling away. The fragment size distribution and PDI showed little dependence on degradation time, which was due to the exponential decrease in the radical initiator over time. After 3 days, the average hydrodynamic size of PS fragments was about 200 nm, decreasing to 20 nm after 9 days, and then remaining constant. This was due to the gradual degradation and peeling of the surface layer, influenced by decreasing $SO_4\bullet^-$ concentration. The core size also decreased exponentially, suggesting mechanisms other than peeling, such as oxygen diffusion limiting degradation. The cracks formed uniformly, and as radicals penetrated, they caused further degradation and embrittlement. The core size rapidly decreased from 11,000 nm to 2500 nm within 6 days, indicating a breakdown mechanism. The SEM images showed irregular spheres with defects, supporting the mechanism. The core size stabilized at 1000 nm after 9 days, maintaining a spherical shape. The degradation caused peeling and cracking, with mechanical stresses leading to the breakdown. The shear forces of seawater caused the PS spheres to peel off uniformly, maintaining their shape while decreasing in diameter.

From an environmental relevance stance, the PS pulverized pellet flakes were used to simulate gradual degradation. The SEM observations showed no shape changes until 24 days of degradation time, but after 30 days, the surface became rough and fragmented. The PS flakes partially broke into pieces around 100 μm after 42 days, influenced by mechanical factors like residual stress and seawater shear stress. When the initial shape was flake-like, the breakdown shape differed greatly from that of the sphere. It was found that the shape at the time of the breakdown was greatly influenced by the shape at the start.

Supplementary Materials: The following supporting information can be downloaded at https://www.mdpi.com/article/10.3390/polym17121715/s1: Figure S1: SEM images of the core parts obtained from PS particles with gradual degradation over various degradation times. Figure S2: FT-IR spectra of gradually degraded PS pulverized pellets at 0 day and 21 days degradation times.

Author Contributions: Methodology, H.N.; formal analysis, H.N., M.A. and M.S.; investigation, H.N.; writing—original draft, H.N. and A.T.N.D.; writing—review and editing, preparation and editing, S.M., H.-J.K., M.Y. and Y.K.; supervision, H.N. All authors have read and agreed to the published version of the manuscript.

Funding: This work was funded by the Environment Research and Technology Development Fund No. 1MF-2204 of the Environmental Restoration and Conservation Agency provided by the Ministry of the Environment of Japan, and by the Grant-in-Aid for Scientific Research No. 23K04822 from Japan Society for the Promotion of Science.

Institutional Review Board Statement: Not applicable.

Data Availability Statement: Data are contained within the article.

Conflicts of Interest: The authors declare no conflicts of interest.

References

1. Thompson, R.C.; Swan, S.H.; Moore, C.J.; vom Saal, F.S. Our plastic age. *Phil. Trans. R. Soc. B* **2009**, *364*, 1973–1976. [CrossRef] [PubMed]
2. Andrady, A.L. Microplastics in the marine environment. *Mar. Pollut. Bull.* **2011**, *62*, 1596–1605. [CrossRef] [PubMed]
3. Jambeck, J.R.; Geyer, R.; Wilcox, C.; Siegler, T.R.; Perryman, M.; Andrady, A.; Narayan, R.; Law, K.L. Plastic waste inputs from land into the ocean. *Science* **2015**, *347*, 768–771. [CrossRef] [PubMed]
4. Law, K.L. Plastics in the Marine Environment. *Annu. Rev. Mar. Sci.* **2017**, *9*, 205–229. [CrossRef]
5. Rillig, M.C.; Kim, S.W.; Kim, T.Y.; Waldman, W.R. The Global Plastic Toxicity Debt. *Environ. Sci. Technol.* **2021**, *55*, 2717–2719. [CrossRef]
6. Avio, C.G.; Gorbi, S.; Regoli, F. Plastics and microplastics in the oceans: From emerging pollutants to emerged threat. *Mar. Environ. Res.* **2017**, *128*, 2–11. [CrossRef]
7. Fauser, P.; Vorkamp, K.; Strand, J. Residual Additives in Marine Microplastics and Their Risk Assessment—A critical review. *Mar. Pollut. Bull.* **2022**, *177*, 113467. [CrossRef]
8. Ter Halle, A.; Jeanneau, L.; Martignac, M.; Jardé, E.; Pedrono, B.; Brach, L.; Gigault, J. Nanoplastic in the North Atlantic Subtropical Gyre. *Environ. Sci. Technol.* **2017**, *51*, 13689–13697. [CrossRef]
9. Pradel, A.; Catrouillet, C.; Gigault, J. The environmental fate of nanoplastics: What we know and what we need to know about aggregation. *NanoImpact* **2023**, *29*, 100453. [CrossRef]
10. Cai, H.; Xu, E.G.; Du, F.; Li, R.; Liu, J.; Shi, H. Analysis of environmental nanoplastics: Progress and challenges. *Chem. Eng. J.* **2021**, *410*, 128208. [CrossRef]
11. Hadri, H.E.; Gigault, J.; Maxit, B.; Grassl, B.; Reynaud, S. Nanoplastic from mechanically degraded primary and secondary microplastics for environmental assessments. *NanoImpact* **2020**, *17*, 100206. [CrossRef]
12. Li, L.; Xu, Y.; Li, S.; Zhang, X.; Feng, H.; Dai, Y.; Zhao, J.; Yue, T. Molecular modeling of nanoplastic transformations in alveolar fluid and impacts on the lung surfactant film. *J. Hazard. Mater.* **2022**, *427*, 127872. [CrossRef] [PubMed]
13. Wan, S.; Wang, X.; Chen, W.; Wang, M.; Zhao, J.; Xu, Z.; Wang, R.; Mi, C.; Zheng, Z.; Zhang, H. Exposure to high dose of polystyrene nanoplastics causes trophoblast cell apoptosis and induces miscarriage. *Part. Fibre Toxicol.* **2024**, *21*, 13. [CrossRef] [PubMed]
14. Marfella, R.; Prattichizzo, F.; Sardu, C.; Fulgenzi, G.; Graciotti, L.; Spadoni, T.; D'Onofrio, N.; Scisciola, L.; La Grotta, R.; Frigé, C.; et al. Microplastics and Nanoplastics in Atheromas and Cardiovascular Events. *N. Engl. J. Med.* **2024**, *390*, 900–910. [CrossRef] [PubMed]
15. Estrela, F.N.; Guimarães, A.T.B.; Araújo, A.P.C.; Silva, F.G.; Luz, T.M.; Silva, A.M.; Pereira, P.S.; Malafaia, G. Toxicity of polystyrene nanoplastics and zinc oxide to mice. *Chemosphere* **2021**, *271*, 129476. [CrossRef]
16. Carlsson, D.J.; Wiles, D.M. The photodegradation of polypropylene films. III. photolysis of polypropylene hydroperoxides. *Macromolecules* **1969**, *6*, 597–606. [CrossRef]
17. Kato, Y.; Carlsson, D.J.; Wiles, D.M. The photo-oxidation of polypropylene: Some effects of molecular order. *J. Appl. Polym. Sci.* **1969**, *13*, 1447–1458. [CrossRef]
18. Lambert, S.; Wagner, M. Formation of microscopic particles during the degradation of different polymers. *Chemosphere* **2016**, *161*, 510–517. [CrossRef]
19. Julienne, F.; Delorme, N.; Lagarde, F. From macroplastics to microplastics: Role of water in the fragmentation of polyethylene. *Chemosphere* **2019**, *236*, 124409. [CrossRef]
20. Gugumus, F. Re-examination of the role of hydroperoxides in polyethylene and polypropylene: Chemical and physical aspects of hydroperoxides in polyethylene. *Polym. Degrad. Stab.* **1995**, *49*, 29–50. [CrossRef]
21. Mikdam, A.; Colina, X.; Minard, G.; Billon, N.; Maurin, R. A kinetic model for predicting the oxidative degradation of additive free polyethylene in bleach desinfected water. *Polym. Degrad. Stab.* **2017**, *146*, 76–94. [CrossRef]
22. Nakatani, H.; Ohshima, Y.; Uchiyama, T.; Motokucho, S.; Dao, A.T.N.; Kim, H.J.; Yagi, M.; Kyozuka, Y. Rapid oxidative fragmentation of polypropylene with pH control in seawater for preparation of realistic reference microplastics. *Sci. Rep.* **2023**, *13*, 4247. [CrossRef] [PubMed]
23. Nakatani, H.; Yamaguchi, T.; Asano, M.; Motokucho, S.; Dao, A.T.N.; Kim, H.J.; Yagi, M.; Kyozuka, Y. Differences in nanoplastic formation behavior between high-density polyethylene and low-density polyethylene. *Molecules* **2025**, *30*, 382. [CrossRef] [PubMed]
24. Li, X.; Tian, Z.; Kong, Y.; Cao, X.; Liu, N.; Zhang, T.; Xiao, Z.; Wang, Z. The suspension stability of nanoplastics in aquatic environments revealed using meta-analysis and machine learning. *J. Hazard. Mater.* **2024**, *471*, 134426. [CrossRef]

25. An, Y.; Kajiwara, T.; Padermshoke, A.; Nguyen, T.V.; Feng, S.; Masunaga, H.; Kobayashi, Y.; Ito, H.; Sasaki, S.; Isobe, A.; et al Photooxidative degradation and fragmentation behaviors of oriented isotactic polypropylene. *Polym. J.* **2024**, *56*, 379. [CrossRef]
26. Julienne, F.; Lagarde, F.; Delorme, N. Influence of the crystalline structure on the fragmentation of weathered polyolefines. *Polym. Degrad. Stab.* **2019**, *170*, 109012. [CrossRef]
27. O'Donnell, B.; White, J.R. Photo-oxidation of polystyrene under load. *J. Mater. Sci.* **1994**, *29*, 3955–3963. [CrossRef]
28. Rabello, M.S.; White, J.R. Crystallization and melting behaviour of photodegraded polypropylene—I. Chemi-crystallization. *Polymer* **1997**, *38*, 6379–6387. [CrossRef]
29. Rabello, M.S.; White, J.R. The role of physical structure and morphology in the photodegradation behaviour of polypropylene. *Polym. Degrad. Stab.* **1997**, *56*, 55–73. [CrossRef]
30. Corcoran, P.L.; Biesinger, M.C.; Grifi, M. Plastics and beaches: A degrading relationship. *Mar. Pollut. Bull.* **2009**, *58*, 80–84. [CrossRef]
31. Cooper, D.A.; Corcoran, P.L. Effects of mechanical and chemical processes on the degradation of plastic beach debris on the island of Kauai, Hawaii. *Mar. Pollut. Bull.* **2010**, *60*, 650–654. [CrossRef] [PubMed]
32. Syranidou, E.; Karkanorachaki, K.; Amorotti, F.; Repouskou, E.; Kroll, K.; Kolvenbach, B.; Corvini, P.F.X.; Fava, F.; Kalogerakis, N. Development of tailored indigenous marine consortia for the degradation of naturally weathered polyethylene films. *PLoS ONE* **2017**, *12*, e0183984. [CrossRef] [PubMed]
33. Zhou, Q.; Zhang, H.; Fu, C.; Zhou, Y.; Dai, Z.; Li, Y.; Tu, C.; Luo, Y. The distribution and morphology of microplastics in coastal soils adjacent to the Bohai Sea and the Yellow Sea. *Geoderma* **2018**, *322*, 201–208. [CrossRef]
34. Wu, X.; Liu, P.; Wang, H.; Huang, H.; Shi, Y.; Yang, C.; Gao, S. Photo aging of polypropylene microplastics in estuary water and coastal seawater: Important role of chlorine ion. *Water Res.* **2021**, *202*, 117396. [CrossRef]
35. Kooi, M.; Nes, E.H.V.; Scheffer, M.; Koelmans, A. Ups and downs in the ocean: Effects of biofouling on vertical transport of microplastics. *Environ. Sci. Technol.* **2017**, *51*, 7963–7971. [CrossRef]
36. Pabortsava, K.; Lampitt, R.S. High concentrations of plastic hidden beneath the surface of the Atlantic Ocean. *Nat. Commun.* **2020**, *11*, 4073. [CrossRef]

polymers MDPI

Article

Sustainable Multi-Cycle Physical Recycling of Expanded Polystyrene Waste for Direct Ink Write 3D Printing and Casting: Analysis of Mechanical Properties

Rubén García-Sobrino †, Alejandro Cortés *,†, José Ignacio Sevilla-García and Marta Muñoz *

Department of Applied Mathematics, Materials Science and Engineering and Electronic Technology, Universidad Rey Juan Carlos, Calle Tulipán s/n, 28933 Móstoles, Spain; ruben.sobrino@urjc.es (R.G.-S.); ji.sevilla.2020@alumnos.urjc.es (J.I.S.-G.)

* Correspondence: alejandro.cortes@urjc.es (A.C.); marta.munoz@urjc.es (M.M.)

† These authors contributed equally to this work.

Abstract: This work investigates the sustainable reuse of expanded polystyrene (EPS) waste through a multi-cycle physical recycling process involving dissolution in acetone and subsequent manufacturing via Direct Ink Write (DIW) 3D printing and casting. Morphology and mechanical properties were evaluated as a function of the manufacturing technique and number of dissolution cycles. Morphological analysis revealed that casted specimens better replicated the target geometry, while voids in 3D-printed specimens aligned with the printing direction due to rapid solvent evaporation. These voids contributed to slightly reduced stiffness in 3D-printed specimens compared to casted ones, particularly for transverse printing orientation. The defoaming process during dissolution significantly increased the density of the material, as well as removed low molecular weight additives like plasticizers, leading to a notable enhancement in stiffness. Successive dissolution cycles led to increased removal of plasticizers, enhancing stiffness up to 52 times (cast), 42 times (longitudinally printed), and 35 times (transversely printed) relative to as-received EPS waste. The glass transition temperature remained unchanged, confirming the preservation of polymer integrity. This work highlights the potential of EPS inks for sustainable, multi-cycle recycling, combining enhanced mechanical performance with the flexibility of 3D printing for complex, cost-effective designs, aligning with circular economy principles.

Keywords: circular economy; plastic waste recycling; expanded polystyrene; additive manufacturing; Direct Ink Write

1. Introduction

The growing environmental impact of expanded polystyrene (EPS) waste has driven the need for innovative recycling methods. The current state of research in EPS recycling highlights several methodologies based on thermal, chemical, and physical methods [1]. Thermal recycling of EPS offers a way to convert waste into energy but requires careful management to be sustainable at both environmental and economic levels in terms of its energy requirements [2–4]. On the other hand, chemical recycling produces new materials that match the quality of conventional virgin materials [5] and may include processes that convert plastic waste into, for instance, carbon nanomaterials [6]. Finally, physical recycling involves physically processing waste EPS into new products without altering its chemical structure.

The latter methodology, i.e., physical recycling, has been prioritized according to the European Commission to minimize environmental impact and promote sustainable waste management practices across its territory [7]. In contrast to chemical recycling, physical recycling has gained attention for its ability to treat more complex plastic waste streams than the former, but it remains an energy-intensive process and is currently not

economically viable on a large scale. In this sense, the European strategy aims to make all plastic packaging on the EU market reusable or recyclable by 2030 [8]. This plan aims to envisage a drastic reduction in the use of single-use plastics, including polystyrene products such as the one cited in this manuscript [9]. Thus, the reduction in plastic packaging is targeted at a percentage decrease of at least 50% by 2025 and 55% by 2030 [9]. These regulations encourage the development and adoption of physical recycling technologies to meet the EU's circular economic goals; however, this technique may be limited by the degradation of material properties over multiple recycling cycles [10].

In this sense, Ferrándiz-Mas et al. [11] evaluated the reuse of EPS to produce lightweight cement mortars containing up to 60% recycled EPS with mechanical properties comparable to cement, making them suitable for specific applications. Some studies explore the dissolution of EPS in acetone, focusing on the kinetics and composition of the resulting swollen polystyrene used as a binder for molding compounds [12]. There are also studies based on the use of more environmentally sustainable solvents such as limonene [13] or natural essences [14].

On the other hand, 3D printing, or additive manufacturing (AM), is emerging as a revolutionary technology in the field of plastics recycling [15–17]. The customization and flexibility that it offers compared to traditional manufacturing technologies, as well as the innovation involved, could facilitate social awareness that promotes the development of a circular economy. To the best of our knowledge, there are several studies in the literature on the recycling of EPS with AM, all of them based on the Fused Filament Fabrication (FFF) technology [18–21]. Briefly, the operation of this technology consists of the extrusion of a thermoplastic type of filament with the addition of temperature. In this way, the extruded material exits towards a bed or deposit through a nozzle, building the final material layer by layer [22]. However, as mentioned above, the temperature required for the aforementioned extrusion process can lead to high energy consumption on an industrial scale. The requirement of a temperature addition conditions the development of a sustainable recycling process.

Unlike the aforementioned works, this manuscript is based on Direct Ink Writing (DIW). This technology is also an extrusion-based 3D printing technology; however, unlike the filament required by the described FFF technology, DIW starts from a viscous paste-based material that is extruded after pressure from a die. To this end, the above-mentioned viscosity allows the printing of elements at or just above room temperature and atmospheric pressure [23,24]. This quality, ideal for our goal, also favors the printing of other materials such as polymers, waxes, hydrogels, ceramics, and even increasing metals [25], which significantly increases the likelihood of recycling, as confirmed by A. Mantelli et al. [26]. In this sense, the work of M. Zheng et al. [27] presents a new type of photothermal polyurethane (PU) that can be recycled and repaired in situ.

In the case of this work, our group was a pioneer in using 3D printing inks based on physically recycled EPS waste for DIW technology [28]. This work presented a novel and accessible method to any user that allows the physical recycling of EPS using acetone as a solvent without damaging the macromolecules [29]. In that work, the use of this EPS-ink for DIW 3D printing was demonstrated and optimized by printing linear circuits suggesting the possibility of an additive manufacturing system minimizing energy and resource consumption, offering sustainable and cost-efficient methods for producing high-value, complex objects. Nevertheless, just one recycling cycle was studied.

Once the efficacy of this technology was validated through simple geometries and a single recycling cycle, the present work aims to extend this analysis by evaluating the feasibility of using this technology for a greater number of recycling cycles, as well as their effect on the mechanical properties. Controversial hypotheses exist regarding the long-term mechanical integrity and environmental impact of recycled EPS products. Some studies indicate polymer chain degradation by a reduction in both molar weight and glass transition temperature, suggesting an oxidative degradation process [30], while others claim that using as dissolution agents for EPS demonstrates no adverse effects on polymer

chain integrity [31]. For this purpose, the effect of conducting three subsequent dissolution and manufacturing cycles with the EPS waste on the mechanical and physicochemical properties was evaluated. Furthermore, the effect of the manufacturing process on the mechanical properties was also studied. Here, the performance of the as-received EPS waste is compared to the one reached by cast and 3D-printed specimens using the physically recycled EPS inks.

2. Materials and Methods

2.1. Materials

In the present work, the EPS waste was obtained from different centers associated with the Rey Juan Carlos University (Madrid, Spain). It is worth mentioning that the EPS waste was washed and dried when it was collected to avoid any contamination. It should also be noted that this material (washed and dried EPS waste) was labeled as "as received" throughout the manuscript. On the other hand, acetone solvent (99.6%) from GLR Labkem was used for defoaming and minimizing the EPS waste, obtaining the EPS inks for the subsequent manufacturing processes described below.

2.2. Preparation of the EPS Inks

The process for obtaining the EPS ink consists of a defoaming process by dissolving the EPS waste in acetone solvent in a 100:1 volume ratio. This process involves the formation of viscous paste suitable for 3D printing using DIW technology. To demonstrate the circularity of the process, three subsequent dissolution and manufacturing cycles were carried out on EPS waste in this research work. Therefore, at this point, it is necessary to define the names of the different EPS products. Here, the as-received EPS waste (EPS-0) is dissolved in acetone to produce EPS-1. Then, after manufacturing and testing specimens with this ink, they were dissolved again in acetone, obtaining EPS-2. Repeating the cycle once more, defoaming and manufacturing yielded EPS-3. Figure 1 illustrates the manufacturing and dissolution processes used to produce the different EPS inks.

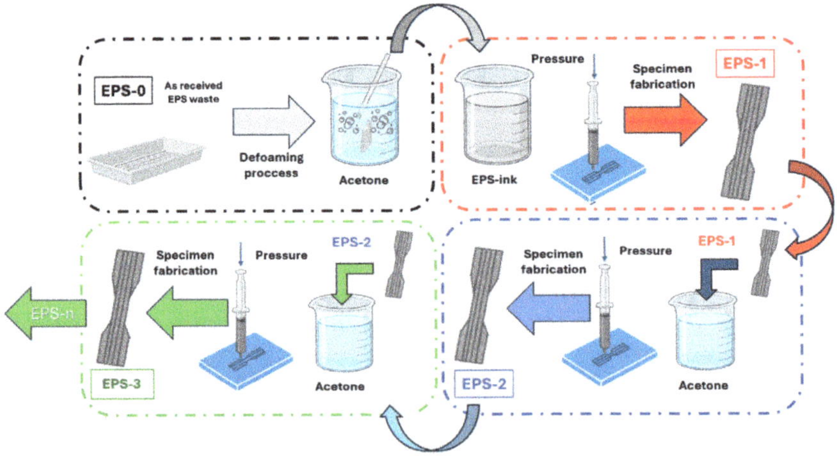

Figure 1. Scheme of the proposed multi-cycle physical recycling process. The as-received EPS waste (EPS-0) was dissolved in acetone, inducing the defoaming process, thus obtaining an EPS-ink (EPS-1). After manufacturing with this ink, the specimens were dissolved again in acetone, obtaining EPS-2, and so on, completing "n" cycles.

2.3. Manufacture of Testing Specimens

In this research work, the effect of the number of dissolution cycles on the mechanical properties was studied as a function of the manufacturing process. Two different manufactur-

ing processes were compared (casting and 3D printing) for the EPS-1, EPS-2, and EPS-3 inks. In addition, the mechanical properties were compared to the as-received EPS waste (EPS-0).

The specimens of the as-received EPS waste (EPS-0) were obtained by CNC machining using a HIGH-Z S-1400 T CNC-Technik equipment associated with the Mach3 Loader software (see Figure 2a). The machining conditions were as follows: spindle speed of 2000 rpm, 2000 feed rate, and 3 mm of pass depth (stepover 0.5 mm), using a milling cutter with a diameter of 4 mm.

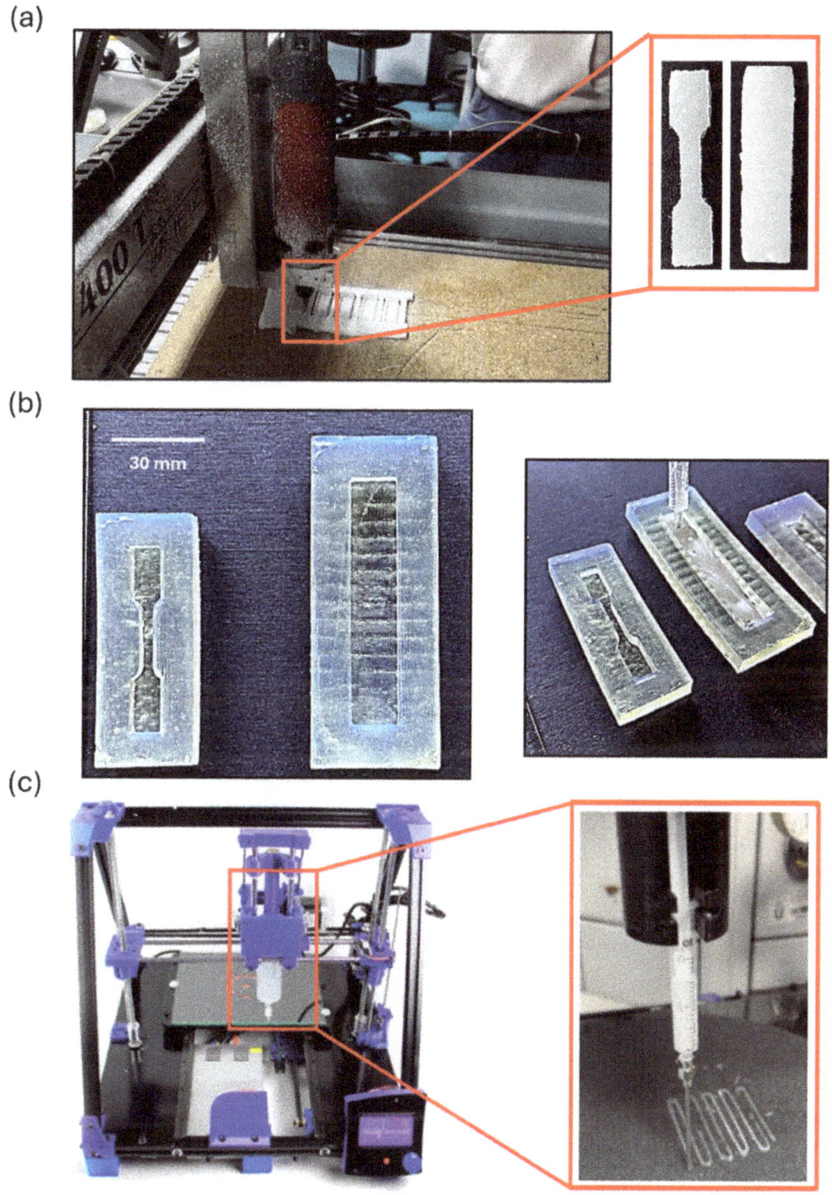

Figure 2. (**a**) CNC machining of the as-received EPS waste (EPS-0) (**b**) Molds obtained by SLA 3D printing technology and casting process using EPS-1, EPS-2, and EPS-3. (**c**) DIW 3D printing technology using a BCN3D Plus printer with a modified Paste Extruder Module.

On the other hand, molds with the desired geometries were designed and printed using stereolithography (SLA) 3D printing technology (Formlabs 3B) in Elastic 50A V1 resin using 100 μm of layer height. After the printing process was completed, the molds were washed with 2-isopropanol (99.9%) supplied by Merck and then post-cured by UV radiation using Form Cure equipment for 40 min at 60 °C. Once the molds were obtained, a casting process was conducted (see Figure 2b) using the EPS-1, EPS-2, and EPS-3 inks.

For the 3D-printed specimens, a BCN3D Plus (Barcelona, Spain) DIW 3D printer, with a modified Paste Extruder module, was used in the present study. The latter element was modified to allow printing from disposable syringes, facilitating the accessibility of the technology presented (see Figure 2c). This added element was also printed with High-Impact Polystyrene (HIPS) filament from SmartFil by Fused Filament Fabrication (FFF) 3D printing technology using a BQ Witbox printer. It is worth mentioning that the specimens were printed on SiC-P120 sandpaper (QATM Quality Assured; $R_a = 24.05 \pm 1.27$ μm) to improve the adhesion with the printed material and minimize the nozzle drag. Most of the 3D printing parameters for the DIW technology, set up using the Slic3r software, were optimized according to a previously published work [28]. More specifically, a 2.11 mm diameter needle (BENECRAFT), a printing speed of 12 mm/s, an extrusion multiplier of 2, a retraction of 0.5 mm, and a printing bed temperature of 25 °C. Furthermore, the nozzle tip was dipped in silicone oil before printing, acting as a lubricant to reduce friction forces. In this research work, the infill percentage was optimized to match the dimensions of the 3D model in two printing orientations, longitudinal and transverse, regarding the longest dimension of specimens. For this task, 15×35 mm^2 rectangular pieces were printed using infill percentages ranging from 10 to 100%, and the morphology of the resulting specimens was evaluated.

All the 3D models for 3D printing were designed using CATIA v5 software (Dassault Systèmes, Velizy-Villacoublay, France).

2.4. Physico-Chemical Characterization

Viscosity characterization was carried out at 25 °C according to ISO-3219 rotational viscosity [32] with a Brookfield DV2T viscometer from AMETEK Brookfield Inc (MA, USA). The spindle used was N-LV3 (63) [33] in the range from 0.1 to 0.4 revolutions per minute (r.p.m.). The viscosity analyses were performed in triplicate.

Weight-average and number-average molecular weights (Mw and Mn, respectively) and the polydispersity index (PDI) were measured by Gel Permeation Chromatography (GPC) with the use of three columns of PL Gel Olexix. The system was calibrated with a narrow molecular weight distribution of standard polystyrene.

Differential Scanning Calorimetry (DSC) tests were carried out using a DSC25 device from TA Instruments (New Castle, DE, USA) from 30 to 200 °C at a 20 °C/min heating rate in a nitrogen atmosphere. The glass transition temperature (Tg) was calculated using the turning point analysis method by TA Instruments TRIOS v5.1.1.46572 software according to ISO-11357 [34].

The density of the samples has been measured by using an analytical balance (Mettler Toledo) with the XPR/SSR-Ana density kit in distilled water at room temperature (20 °C) following the UNE-EN ISO 1183-1 standard [35]. The Archimedes density method was employed to assess the density of the recycled EPS under study. The results were compared with the density obtained for the as-received EPS waste. Each specimen was weighed in air and then submerged in distilled water to calculate the Archimedes density (*D*), using the Equation (1):

$$D = \frac{A}{A-B}(D_0 - D_l) + D_l \tag{1}$$

where (*A*) represents the weight of the sample measured in air; (D_0) is the density of the distilled water in which the part is submerged, taking its temperature into account; (*Dl*) is the air density under room conditions with a value of 0.0012 g/mL; and (*B*) is the weight of the sample when immersed in water.

2.5. Mechanical Properties

The mechanical properties were analyzed in tensile and 3-point bending load conditions. In both cases, the universal testing machine used was a Zwick Z100 equipment from ZwickRoell, using a load cell of 5 kN. The dimensions and calculations were conducted according to ASTM D638 [36] for the tensile tests, which were carried out at 5 mm/s speed, whereas the 3-point bending tests were evaluated according to ASTM D790 [37] with a test speed of 10 mm/s. Finally, it should be noted that all studies were conducted in quintuplicate.

3. Results and Discussion

3.1. Optimization of the 3D Printing Process

Optimization of the infill percentage of the 3D printing process was conducted to match the dimensions of the 3D model in the two proposed printing orientations (longitudinal and transverse). Figure 3a,b show representative images of the specimens used for the infill percentage optimization study, manufactured in the longitudinal and transverse printing orientations, respectively.

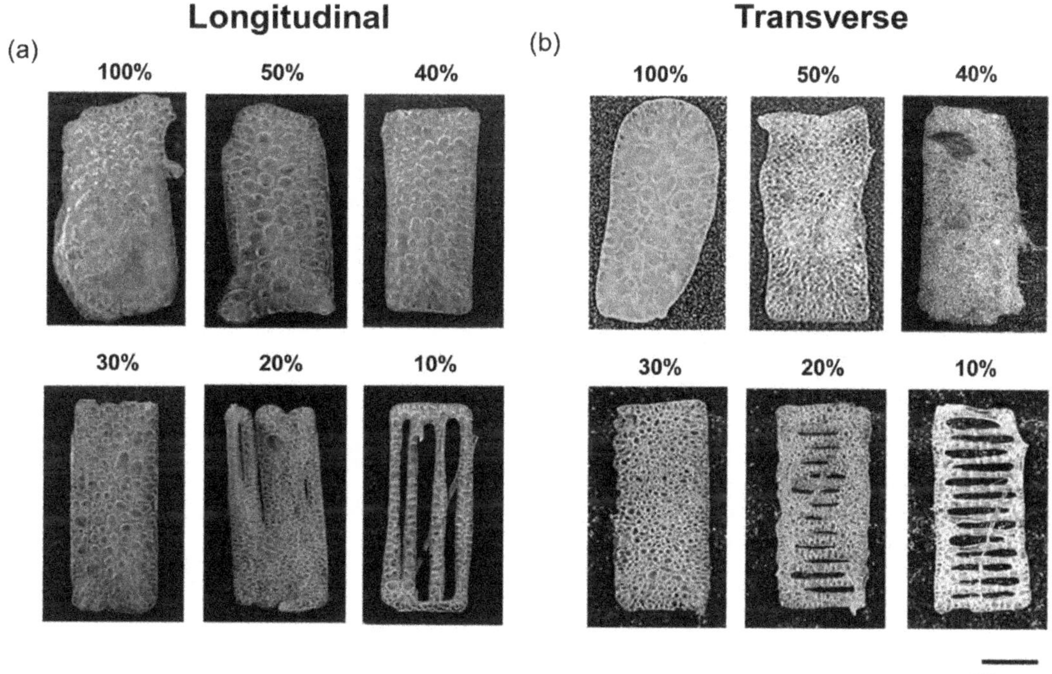

Figure 3. Representative images of the specimens used for the infill percentage optimization study, manufactured in the longitudinal (**a**) and transverse (**b**) printing orientations.

In both cases, a high infill percentage results in geometric distortions relative to the 3D model due to an excess of extruded volume, as clearly observed in specimens with 50% and 100% infill percentages. Conversely, a low infill percentage leads to insufficient filling, resulting in discontinuous specimens, as seen in those with 10% and 20% infill.

Table 1 shows the quantitative dimensional study by measuring and comparing the dimensions of the 3D-printed specimens to the 3D model carried out by image analysis using the continuous specimens (infill percentage above 20%). As previously mentioned, the increase in the infill percentage led to a higher width and length caused by the excess

of extruded material. Furthermore, the excess of extruded material yielded a higher geometrical distortion, which was calculated as the standard deviation in length and width. At this point, it is important to highlight the significant presence of voids, regardless of the printing orientation or infill percentage. As demonstrated in our previous research work, these voids are the result of the evaporation of acetone, whose vapor is trapped in the printed sample due to the formation of a thin, dry film on its surface [28]. The kinetics of this process depend on the volume of material expelled [38].

Table 1. Quantitative dimensional study.

		3D Model			
	Width (mm)			15	
	Length (mm)			35	
		3D-printed—Longitudinal			
	Infill (%)	100	50	40	30
Width	Average width (mm)	16.32 ± 1.66	16.15 ± 1.39	15.55 ± 0.57	15.61 ± 0.32
	Width deviation (%)	10.17	8.65	3.66	2.04
Length	Average length (mm)	35.97 ± 2.16	35.46 ± 0.96	35.83 ± 0.41	35.76 ± 0.28
	Length deviation (%)	6.01	2.71	1.14	0.78
		3D-printed—Transverse			
	Infill (%)	100	50	40	30
Width	Average width (mm)	14.87 ± 1.88	15.92 ± 0.67	15.64 ± 0.90	15.84 ± 0.27
	Width deviation (%)	12.64	6.17	5.75	1.70
Length	Average length (mm)	36.02 ± 3.66	35.77 ± 1.05	35.86 ± 1.15	35.61 ± 0.44
	Length deviation (%)	10.16	2.93	3.2	1.23

In summary, considering that a higher infill percentage leads to an excess of extruded material and an increased acetone vapor void formation, which promotes geometric distortions in the specimens compared to the 3D model, the lowest infill percentage that produced a continuous specimen, 30%, was selected as the optimum.

3.2. Morphological Analysis

Once the repeatability of the 3D printing technique for obtaining solid two-dimensional systems was optimized, the mechanical properties were evaluated as a function of the manufacturing technique and the number of dissolution cycles in acetone. In addition, the influence of the layer orientation of the 3D-printed specimens on the mechanical properties was also studied.

First, Figure 4 shows the parts obtained with the different manufacturing processes and the number of dissolution cycles. Here, it is worth mentioning the noticeable change in the look and physical properties of the manufactured parts when carrying out the defoaming process (EPS-1) from the as-received EPS waste (EPS-0). Furthermore, the repeatability of the manufacturing processes when conducting the subsequent dissolution cycles is noteworthy, showing the specimens similar dimensions and looks, regardless of the number of dissolution cycles. This illustrates the sustainability of the proposed material since it is possible to recycle the parts several times when reaching their end-of-life, as it is a physical recycling process [31].

When analyzed in detail, the casted specimens replicate the desired geometry better than the 3D-printed ones, which can be ascribed to the inherent nature of DIW 3D printing [39]. Unlike casting, there are no molds in DIW, resulting in slight distortions in the specimen walls. In addition, the syringe drags part of the extruded material nearby when printing, leading to shape distortions, especially when sudden changes in the printing direction occur.

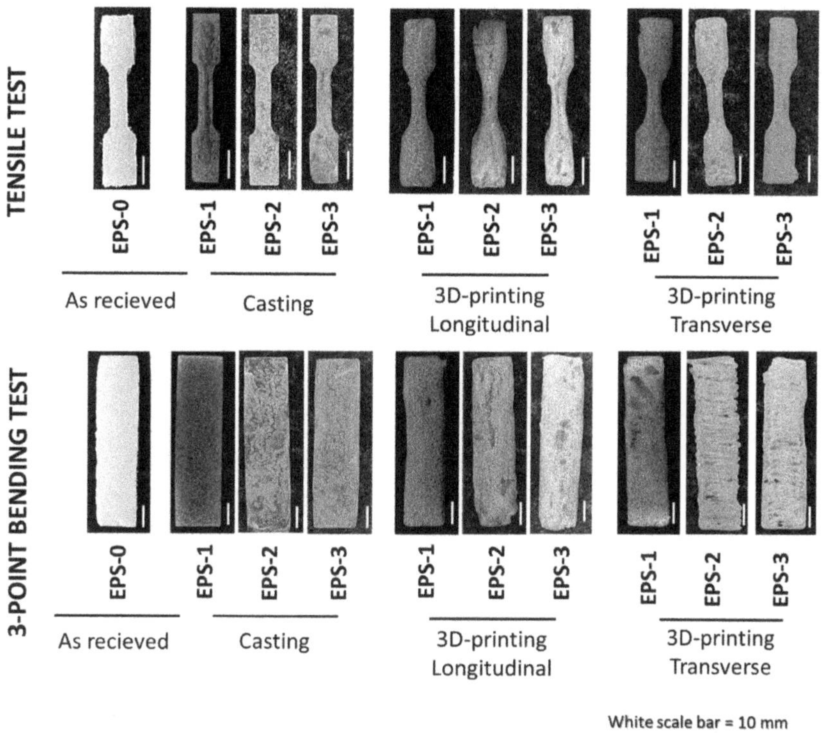

Figure 4. Parts obtained for the tensile and the 3-point bending tests with the different manufacturing methods: CNC machining of as-received EPS waste (EPS-0), casting (EPS-1 to 3), and 3D printing (EPS-1 to 3) in the two proposed printing orientations, longitudinal and transverse.

Finally, the void distribution is different between the manufacturing conditions. More specifically, the voids are randomly distributed in the casted specimens, whereas the 3D-printed ones present aligned porosity in the direction of the layer orientation. This can be explained due to the rapid evaporation of the solvent. In contrast to casting, where the viscous paste fills the mold uniformly, the volume of the printed part is filled by extruding continuous parallel lines in DIW 3D printing. The solvent evaporates almost instantly from the printed ribbon surface during printing, generating a thin, solid skin that impedes ink flow with the subsequent parallel ribbon. Therefore, both the possible entrapped air between ribbons and the voids grown inside each printed ribbon would remain aligned in the printing direction. The effect of the aforementioned features on the mechanical properties will be further discussed.

3.3. Mechanical Properties as a Function of the Manufacturing Technology

Figure 5a,b show representative stress–strain curves obtained from the tensile and 3-point bending tests, respectively, as a function of the manufacturing process, where the as-received specimens were obtained using the EPS-0, and the casting and 3D-printed specimens were obtained using EPS-1, after conducting the defoaming process by dissolution in acetone solvent. Here, all specimens obtained with EPS-1 exhibited significantly increased mechanical properties compared to the received EPS waste, demonstrating higher stiffness and tensile strength, despite suffering a marked decrease in ductility, particularly under tensile loading.

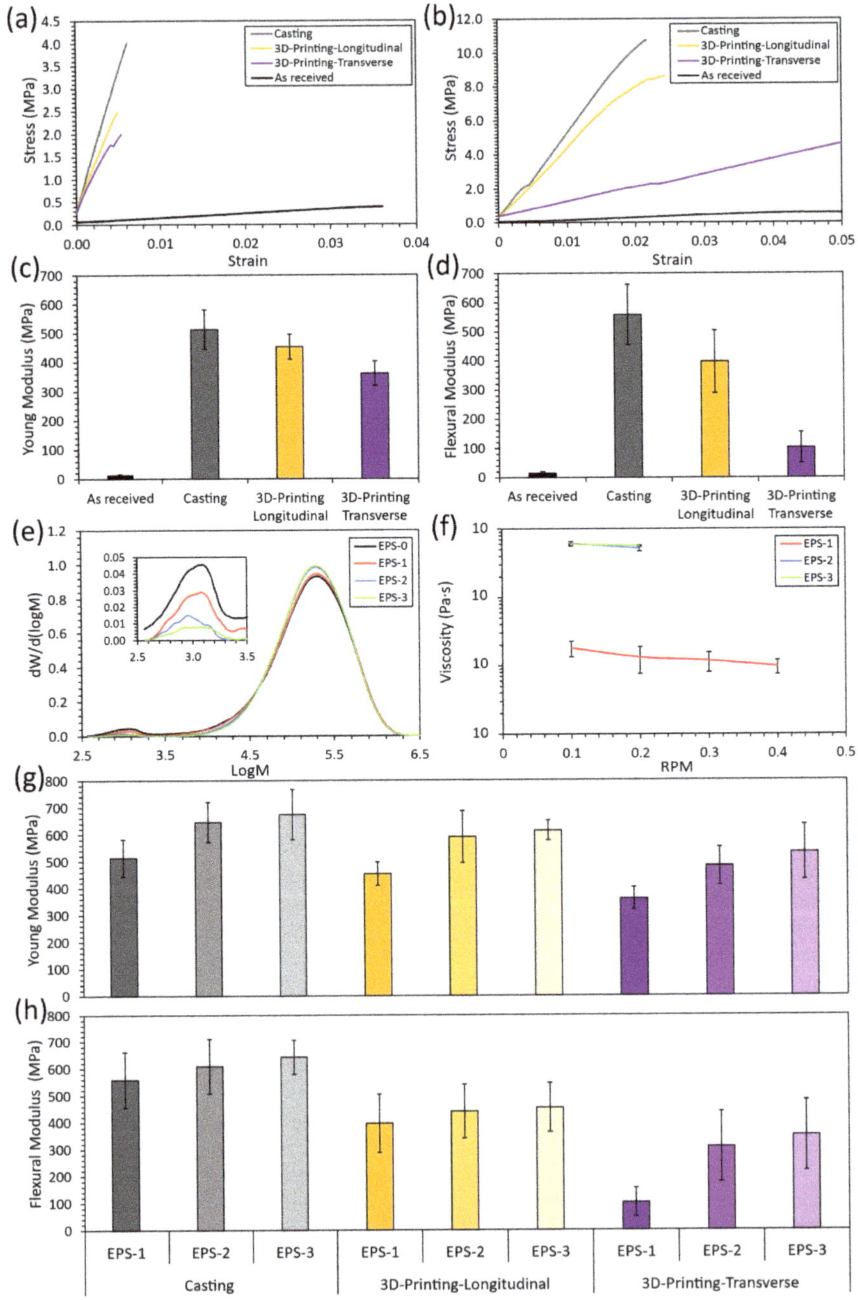

Figure 5. Example curves of tensile (**a**) and three-point bending tests (**b**). Young's (**c**) and flexural (**d**) modulus as a function of manufacturing technology, EPS-0 for the as-received material and EPS-1 for the casted and 3D-printed specimens. Molecular weight distribution obtained by GPC (**e**) and viscosity (**f**) as a function of the number of dissolution cycles in acetone. Young's (**g**) and flexural (**h**) modulus as a function of manufacturing technology and the number of dissolution cycles in acetone.

More specifically, Figure 5c,d show an in-depth analysis of stiffness, which is the most useful property when comparing brittle materials [40] for the tensile and 3-point bending tests, respectively [41,42].

As previously mentioned, a significant increase in stiffness is shown by the casted and 3D-printed specimens. These results can be ascribed to the defoaming process of the as-received EPS waste, which increases density and removes some additives, such as plasticizers. The increase in density when dissolving the as-received EPS waste in acetone is shown in Table 2. Here, the density increases around one order of magnitude, confirming that the defoaming process has taken place. In addition, no considerable differences were observed when increasing the number of dissolution cycles, evincing that the defoaming process is completed from the first dissolution cycle. When comparing the density of the physically recycled specimens with the density of bulk polystyrene (0.96–1.05 g/mL), a lower density is obtained, which can be attributed to the void growth caused by solvent evaporation. The removal of additives, such as plasticizers, was demonstrated in previous research work [28] and evinced again in this work by GPC analysis (see Figure 5e). Here, the decrease in the peak ranging from 2.5 to 3.5 logM indicates the removal of a component with low molecular weight, which was associated with the removal of additives. This point will be further discussed.

Table 2. Density study as a function of the number of dissolution cycles in acetone.

Sample	Density (g/mL)
EPS-0	0.05 ± 0.03
EPS-1	0.52 ± 0.03
EPS-2	0.51 ± 0.03
EPS-3	0.52 ± 0.02

However, analyzing the mechanical properties in detail, the decrease in the mechanical properties of the 3D-printed specimens is also noticeable concerning the casting ones, which can be explained by the internal structure of the specimens. As previously mentioned in Section 3.2, the ink flow between adjacent ribbons in the 3D-printed specimens is hindered due to the rapid formation of a thin, dry film on each ribbon's surface when extruded, leading to a lack of cohesion and entrapped air between adjacent ribbons. Finally, when comparing the two different printing orientations, the specimens printed in the transverse orientation exhibited a lower stiffness than the ones printed in the longitudinal orientation. These results support the assumptions about the rapid formation of the thin, dry film on the ribbon's surface when extruded since the specimens printed in the transverse orientation present more parallel ribbons, leading to the previously mentioned lack of cohesion and entrapped air between adjacent ribbons. Thus, the manufacturing defects are, in this case, aligned perpendicular to the applied forces both in the tensile and three-point bending tests, resulting in a decrease in the mechanical properties [43,44].

It is worth noting that, despite the reduction in the mechanical properties of the 3D-printed specimens compared to the casted ones, stiffness increased on average by more than 36 times for those printed in the longitudinal orientation and over 28 times for those printed in the transverse orientation, relative to the as-received EPS waste (average values obtained using the EPS-1 ink under tensile and three-point bending load conditions). Furthermore, this increase in stiffness, combined with the advantages of 3D printing, enables the fabrication of more complex or customized parts, including unique designs, without significantly increasing production costs. Unlike traditional manufacturing methods, such as casting, 3D printing does not require molds that need to be amortized, making it a more flexible and cost-effective solution for producing intricate geometries or tailored components.

3.4. Mechanical Properties as a Function of the Number of Dissolution Cycles

To study the mechanical properties as a function of the number of dissolution cycles, it is essential to first examine how these cycles affect the properties of the EPS ink. Figure 5e shows the GPC analysis as a function of the dissolution cycles in acetone, whereas Table 3 summarizes the main results obtained from GPC and DSC analyses. Here, the decrease in the peak ranging from 2.5 to 3.5 logM when increasing the number of dissolution cycles is again related to the remotion of polymer additives, such as plasticizers, presenting low molecular weight. This is also directly related to the decrease in the polydispersity index (PDI) and the increase in Mw and Mn, pointing out that the polymer chain length distribution becomes more homogeneous when increasing the number of dissolution cycles. Furthermore, the DSC analysis shows that the glass transition temperature is not affected by the number of dissolution cycles, according to our previous study, since this is a physical recycling process that does not cause damage to the polymer chains as no chemical reactions are involved [28,29].

Table 3. GPC and DSC analysis as a function of the number of dissolution cycles in acetone.

Value/Label	EPS-0	EPS-1	EPS-2	EPS-3
Mw (g/mol) $\times 10^4$	23.85	24.46	24.26	24.75
Mn (g/mol) $\times 10^4$	2.77	4.25	6.42	7.66
PDI	8.61	5.75	3.78	3.23
Tg (°C)	103.08	102.95	104.39	104.76

However, the removal of polymer additives when increasing the number of dissolution cycles significantly increases ink viscosity, as shown in Figure 5f, which could hinder the manufacturing process. Despite this, both the removal of additives and the increase in viscosity tend to stabilize from the second dissolution cycle (EPS-2) onward. Furthermore, the increase in the number of dissolution cycles enhances stiffness in both tensile and three-point bending load conditions, as shown in Figure 5g,h, respectively, again related to the remotion of additives, such as plasticizers. More specifically, the stiffness increased on average by more than 52 times for the cast specimens, over 42 times for those 3D-printed in the longitudinal orientation, and over 35 times for those printed in the transverse orientation, in comparison to the as-received EPS waste (average values obtained using the EPS-3 ink under tensile and three-point bending load conditions).

In this sense, considering that this maximum of viscosity, reached in the third dissolution cycle (EPS-3), does not impede manufacturing by casting and is within the printability limits of DIW 3D printing, no limitations in the number of dissolution cycles are expected, evincing the sustainability of the proposed inks based on EPS waste.

Furthermore, it is worth highlighting the projection of the proposed technology in terms of circular economy and sustainability. In addition to significantly reducing the volume of EPS waste with a minimum amount of solvent (100:1 volume ratio) while enhancing mechanical properties, the process is fully scalable. In fact, any company or home user that has access to a conventional FFF 3D printing machine can freely download files to cheaply manufacture and assemble a DIW device to conduct the proposed recycling process, thus giving a new useful life to EPS waste and obtaining on-demand custom or even unique parts considering the advantages of additive manufacturing. Likewise, acetone is a common solvent, affordable, and easy to acquire, which further facilitates its scalability.

4. Conclusions

The present research work focuses on optimizing the reuse of expanded polystyrene waste (EPS-0) through a physical recycling process involving dissolution in acetone and subsequent manufacturing by two methods: DIW 3D printing and casting. This study explores the influence of 3D printing parameters, such as infill percentage and printing

orientation, on the morphological and mechanical properties of the fabricated specimens. Additionally, the impact of multiple recycling cycles on ink properties, including viscosity and molecular structure, as well as the mechanical performance of the resulting components, was evaluated. The ink obtained in the first dissolution cycle was named EPS-1, whereas those obtained in the second and third dissolution cycles were named EPS-2 and EPS-3, respectively.

Regarding the optimization of 3D printing parameters, the optimal infill percentage for 3D printing was determined to be 30%, as it provided continuous specimens without geometric distortions or insufficient filling. A higher infill percentage ($\geq$50%) caused geometric distortions due to excessive extruded material, while a lower infill percentage ($\leq$20%) resulted in discontinuities. Then, the morphological analysis showed that the casted specimens replicated the desired geometry more accurately than the optimized 3D-printed ones, likely due to the absence of molds in 3D printing and material dragging during deposition. Voids were randomly distributed in cast specimens but aligned with the printing direction in 3D-printed specimens, stemming from the rapid formation of a thin, dry film on the extruded ribbon's surface during printing, hindering ink flow between adjacent printed ribbons.

This fact explains the slight decrease in mechanical properties of the 3D-printed specimens compared to the casted ones, especially for the ones printed in the transverse orientation due to the alignment of defects perpendicular to the applied forces in the latter. The increase in the number of dissolution cycles resulted in a higher removal of low molecular weight additives like plasticizers, leading to increased stiffness. The glass transition temperature remained unaffected (around 104 °C), confirming that the physical recycling process preserved polymer integrity. Stiffness improved with successive dissolution cycles, achieving a maximum enhancement of 52 times (casted), 42 times (longitudinally printed), and 35 times (transversely printed) compared to as-received EPS waste (12.5 $\pm$ 3.2 MPa in tensile load conditions) due to the increase in density (in a factor of 10) related to untreated EPS waste (0.05 $\pm$ 0.03 g/mL), caused by the defoaming process.

Despite the decrease in stiffness of 3D-printed parts compared to those casted, 3D printing offers advantages over casting by enabling the fabrication of complex or customized parts without molds, reducing production costs, especially for intricate designs.

Viscosity increased with additional cycles, stabilizing from the second cycle (EPS-2) onward without impeding manufacturability. Therefore, the physical recycling process demonstrates a sustainable approach to reusing EPS waste, allowing for multiple life cycles of the material while significantly enhancing mechanical properties.

Author Contributions: Conceptualization, R.G.-S., A.C. and M.M.; methodology, R.G.-S., A.C., J.I.S.-G. and M.M.; validation, R.G.-S., A.C. and M.M.; formal analysis, R.G.-S., A.C. and M.M.; investigation, R.G.-S., A.C., J.I.S.-G. and M.M.; data curation, R.G.-S., A.C. and M.M.; writing—original draft preparation, R.G.-S., A.C. and M.M.; writing—review and editing, R.G.-S., A.C. and M.M.; visualization, R.G.-S., A.C. and M.M.; supervision, R.G.-S., A.C. and M.M.; project administration, M.M.; funding acquisition, M.M. All authors have read and agreed to the published version of the manuscript.

Funding: The publication is part of the TED2021-129688B project, funded by MCIN/AEI/10.13039/501100011033 and by the European Union "NextGenerationEU"/PRTR. The content of this publication does not reflect the official opinion of the European Union. Responsibility for the information and views expressed in this paper lies entirely with the authors.

Institutional Review Board Statement: Not applicable.

Data Availability Statement: Data are available upon request.

Conflicts of Interest: The authors declare no conflicts of interest. The funders had no role in the design of the study, in the collection, analysis, or interpretation of data, in the writing of the manuscript, or in the decision to publish the results.

References

1. Netsch, N.; Simons, M.; Feil, A.; Leibold, H.; Richter, F.; Slama, J.; Yogish, S.P.; Greiff, K.; Stapf, D. Recycling of Polystyrene-Based External Thermal Insulation Composite Systems—Application of Combined Mechanical and Chemical Recycling. *Waste Manag.* **2022**, *150*, 141–150. [CrossRef] [PubMed]
2. Arandes, J.M.; Ereña, J.; Azkoiti, M.J.; Olazar, M.; Bilbao, J. Thermal Recycling of Polystyrene and Polystyrene-Butadiene Dissolved in a Light Cycle Oil. *J. Anal. Appl. Pyrolysis* **2003**, *70*, 747–760. [CrossRef]
3. Armenise, S.; SyieLuing, W.; Ramírez-Velásquez, J.M.; Launay, F.; Wuebben, D.; Ngadi, N.; Rams, J.; Muñoz, M. Plastic Waste Recycling via Pyrolysis: A Bibliometric Survey and Literature Review. *J. Anal. Appl. Pyrolysis* **2021**, *158*, 105265. [CrossRef]
4. Wong, S.L.; Armenise, S.; Nyakuma, B.B.; Bogush, A.; Towers, S.; Lee, C.H.; Wong, K.Y.; Lee, T.H.; Rebrov, E.; Muñoz, M. Plastic Pyrolysis over HZSM-5 Zeolite and Fluid Catalytic Cracking Catalyst under Ultra-Fast Heating. *J. Anal. Appl. Pyrolysis* **2023**, *169*, 105793. [CrossRef]
5. Schäfer, P.; Hopmann, C.; Facklam, M.; Hollerbach, L.; Kolb, T.; Schedl, A.; Schmidt, H.-W.; Nosić, F.; Wilhelmus, B. Continuous Chemical Recycling of Polystyrene with a Twin–Screw Extruder. In *Advances in Polymer Processing 2020*; Springer: Berlin/Heidelberg, Germany, 2020; pp. 37–49.
6. Wong, S.L.; Mong, G.R.; Nyakuma, B.B.; Ngadi, N.; Wong, K.Y.; Hernández, M.M.; Armenise, S.; Chong, C.T. Upcycling of Plastic Waste to Carbon Nanomaterials: A Bibliometric Analysis (2000–2019). *Clean Technol. Environ. Policy* **2022**, *24*, 739–759. [CrossRef]
7. Parlamento Europeo y del Consejo. Directiva (UE) 2019-904 de 5 de Junio de 2019, Relativa a La Reducción Del Impacto de Determinados Productos de Plástico En El Medio Ambiente. In *Diario Oficial de la Unión Europea*; European Union: Maastricht, The Netherlands, 2019; pp. 1–19.
8. European Plastics Strategy: Adopted in 2018. Available online: https://environment.ec.europa.eu/strategy/plastics-strategy_en (accessed on 1 November 2024).
9. Single-Use Plastics Directive (EU) 2019/904. Available online: https://eur-lex.europa.eu/eli/dir/2019/904/oj (accessed on 1 November 2024).
10. Schyns, Z.O.G.; Shaver, M.P. Mechanical Recycling of Packaging Plastics: A Review. *Macromol. Rapid Commun.* **2021**, *42*, 2000415. [CrossRef]
11. Ferrándiz-Mas, V.; Sarabia, L.A.; Ortiz, M.C.; Cheeseman, C.R.; García-Alcocel, E. Design of Bespoke Lightweight Cement Mortars Containing Waste Expanded Polystyrene by Experimental Statistical Methods. *Mater. Des.* **2016**, *89*, 901–912. [CrossRef]
12. Ponomarenko, O.; Yevtushenko, N.; Lysenko, T.; Solonenko, L.; Shynsky, V. A New Technology for Producing the Polystyrene Foam Molds Including Implants at Foundry Industry. In *Design, Simulation, Manufacturing: The Innovation Exchange*; Springer: Cham, Switzerland, 2020; pp. 430–437.
13. Kol, R.; Denolf, R.; Bernaert, G.; Manhaeghe, D.; Bar-Ziv, E.; Huber, G.W.; Niessner, N.; Verswyvel, M.; Lemonidou, A.; Achilias, D.S.; et al. Increasing the Dissolution Rate of Polystyrene Waste in Solvent-Based Recycling. *ACS Sustain. Chem. Eng.* **2024**, *12*, 4619–4630. [CrossRef]
14. Gil-Jasso, N.D.; Giles-Mazón, E.A.; Soriano-Giles, G.; Reinheimer, E.W.; Varela-Guerrero, V.; Ballesteros-Rivas, M.F. A Methodology for Recycling Waste Expanded Polystyrene Using Flower Essential Oils. *Fuel* **2022**, *307*, 121835. [CrossRef]
15. Kassab, A.; Al Nabhani, D.; Mohanty, P.; Pannier, C.; Ayoub, G.Y. Advancing Plastic Recycling: Challenges and Opportunities in the Integration of 3D Printing and Distributed Recycling for a Circular Economy. *Polymers* **2023**, *15*, 3881. [CrossRef]
16. Olawumi, M.A.; Oladapo, B.I.; Ikumapayi, O.M.; Akinyoola, J.O. Waste to Wonder to Explore Possibilities with Recycled Materials in 3D Printing. *Sci. Total Environ.* **2023**, *905*, 167109. [CrossRef] [PubMed]
17. Hassan, M.; Mohanty, A.K.; Misra, M. 3D Printing in Upcycling Plastic and Biomass Waste to Sustainable Polymer Blends and Composites: A Review. *Mater. Des.* **2024**, *237*, 112558. [CrossRef]
18. Turku, I.; Kasala, S.; Kärki, T. Characterization of Polystyrene Wastes as Potential Extruded Feedstock Filament for 3D Printing. *Recycling* **2018**, *3*, 57. [CrossRef]
19. Bartolomei, S.S.; da Silva, F.L.F.; de Moura, E.A.B.; Wiebeck, H. Recycling Expanded Polystyrene with a Biodegradable Solvent to Manufacture 3D Printed Prototypes and Finishing Materials for Construction. *J. Polym. Environ.* **2022**, *30*, 3701–3717. [CrossRef]
20. Chu, J.S.; Koay, S.C.; Chan, M.Y.; Choo, H.L.; Ong, T.K. Recycled Plastic Filament Made from Post-consumer Expanded Polystyrene and Polypropylene for Fused Filamant Fabrication. *Polym. Eng. Sci.* **2022**, *62*, 3786–3795. [CrossRef]
21. William, L.J.W.; Koay, S.C.; Chan, M.Y.; Pang, M.M.; Ong, T.K.; Tshai, K.Y. Recycling Polymer Blend Made from Post-Used Styrofoam and Polypropylene for Fuse Deposition Modelling. *J. Phys. Conf. Ser.* **2021**, *2120*, 012020. [CrossRef]
22. González-Henríquez, C.M.; Sarabia-Vallejos, M.A.; Rodriguez-Hernandez, J. Polymers for Additive Manufacturing and 4D-Printing: Materials, Methodologies, and Biomedical Applications. *Prog. Polym. Sci.* **2019**, *94*, 57–116. [CrossRef]
23. Zhu, C.; Liu, T.; Qian, F.; Chen, W.; Chandrasekaran, S.; Yao, B.; Song, Y.; Duoss, E.B.; Kuntz, J.D.; Spadaccini, C.M.; et al. 3D Printed Functional Nanomaterials for Electrochemical Energy Storage. *Nano Today* **2017**, *15*, 107–120. [CrossRef]
24. Li, L.; Lin, Q.; Tang, M.; Duncan, A.J.E.; Ke, C. Advanced Polymer Designs for Direct-Ink-Write 3D Printing. *Chem. A Eur. J.* **2019**, *25*, 10768–10781. [CrossRef]
25. Lewis, J.A. Direct Ink Writing of 3D Functional Materials. *Adv. Funct. Mater.* **2006**, *16*, 2193–2204. [CrossRef]
26. Mantelli, A.; Romani, A.; Suriano, R.; Levi, M.; Turri, S. Direct Ink Writing of Recycled Composites with Complex Shapes: Process Parameters and Ink Optimization. *Adv. Eng. Mater.* **2021**, *23*, 2100116. [CrossRef]

27. Zheng, M.; Guo, Q.; Yin, X.; Getangama, N.N.; de Bruyn, J.R.; Xiao, J.; Bai, Y.; Liu, M.; Yang, J. Direct Ink Writing of Recyclable and in Situ Repairable Photothermal Polyurethane for Sustainable 3D Printing Development. *J. Mater. Chem. A Mater.* **2021**, *9*, 6981–6992. [CrossRef]
28. García-Sobrino, R.; Cortés, A.; Calderón-Villajos, R.; Díaz, J.G.; Muñoz, M. Novel and Accessible Physical Recycling for Expanded Polystyrene Waste with the Use of Acetone as a Solvent and Additive Manufacturing (Direct Ink-Write 3D Printing). *Polymers* **2023**, *15*, 3888. [CrossRef] [PubMed]
29. de Sousa Cunha, R.; Mumbach, G.D.; Machado, R.A.F.; Bolzan, A. A Comprehensive Investigation of Waste Expanded Polystyrene Recycling by Dissolution Technique Combined with Nanoprecipitation. *Environ. Nanotechnol. Monit. Manag.* **2021**, *16*, 100470. [CrossRef]
30. Cella, R.F.; Mumbach, G.D.; Andrade, K.L.; Oliveira, P.; Marangoni, C.; Bolzan, A.; Bernard, S.; Machado, R.A.F. Polystyrene Recycling Processes by Dissolution in Ethyl Acetate. *J. Appl. Polym. Sci.* **2018**, *135*, 46208. [CrossRef]
31. Maharana, T.; Negi, Y.S.; Mohanty, B. Review Article: Recycling of Polystyrene. *Polym. Plast. Technol. Eng.* **2007**, *46*, 729–736. [CrossRef]
32. *ISO-3219*; Olymers/Resins in the Liquid State or as Emulsions or Dispersions—Determination of Viscosity Using a Rotational Viscometer with Defined Shear Rate. ISO: Geneva, Switzerland, 1993. Available online: https://www.iso.org/obp/ui/#iso:std:iso:3219:ed-2:v1:en (accessed on 1 November 2024).
33. AMETEK Brookfield Inc. *More Solutions to Sticky Problems*; AMETEK Brookfield Inc.: Middleborough, MA, USA, 2017; pp. 1–62.
34. *ISO-11357*; Plastics—Differential Scanning Calorimetry (DSC). ISO: Geneva, Switzerland, 2016. Available online: https://www.iso.org/standard/70024.html (accessed on 1 November 2024).
35. *ISO 1183-1:2019*; Plastics: Methods for Determining the Density of Non-Cellular Plastics. ISO: Geneva, Switzerland, 2019. Available online: https://www.iso.org/standard/74990.html (accessed on 1 November 2024).
36. *ASTM D638-22*; Standard Test Method for Tensile Properties of Plastics. ASTM International: West Conshohocken, PA, USA, 2022. Available online: https://www.astm.org/standards/d638 (accessed on 1 November 2024).
37. *ASTM D790-17*; Standard Test Methods for Flexural Properties of Unreinforced and Reinforced Plastics and Electrical Insulating Materials. ASTM International: West Conshohocken, PA, USA, 2017. Available online: https://www.astm.org/standards/d790 (accessed on 1 November 2024).
38. García-Sobrino, R.; Lago, E.; Goñi, C.; Ramos, V.; García, C.; Reinecke, H.; Elvira, C.; Rodríguez-Hernández, J.; Gallardo, A.; Martínez-Campos, E. Fabrication of 3D Cylindrical Thermosensitive Hydrogels as Supports for Cell Culture and Detachment of Tubular Cell Sheets. *Biomater. Adv.* **2023**, *144*, 213210. [CrossRef]
39. Saleh Alghamdi, S.; John, S.; Roy Choudhury, N.; Dutta, N.K. Additive Manufacturing of Polymer Materials: Progress, Promise and Challenges. *Polymers* **2021**, *13*, 753. [CrossRef]
40. Cortés, A.; Sánchez-Romate, X.F.; Jiménez-Suárez, A.; Campo, M.; Ureña, A.; Prolongo, S.G. Mechanical and Strain-Sensing Capabilities of Carbon Nanotube Reinforced Composites by Digital Light Processing 3D Printing Technology. *Polymers* **2020**, *12*, 975. [CrossRef]
41. Wang, X.; Zhao, L.; Fuh, J.Y.H.; Lee, H.P. Effect of Porosity on Mechanical Properties of 3D Printed Polymers: Experiments and Micromechanical Modeling Based on X-Ray Computed Tomography Analysis. *Polymers* **2019**, *11*, 1154. [CrossRef]
42. Jhunjhunwala, P.; Gupta, A. Effect of Porosity on the Quality of 3D Printed Structures. *Int. J. Adv. Manuf. Technol.* **2023**, *127*, 899–909. [CrossRef]
43. Zohdi, N.; Yang, R. (Chunhui) Material Anisotropy in Additively Manufactured Polymers and Polymer Composites: A Review. *Polymers* **2021**, *13*, 3368. [CrossRef] [PubMed]
44. Nath, S.D.; Nilufar, S. An Overview of Additive Manufacturing of Polymers and Associated Composites. *Polymers* **2020**, *12*, 2719. [CrossRef] [PubMed]

polymers

MDPI

Article

Biodegradation Assessment of Bioplastic Carrier Bags Under Industrial-Scale Composting Conditions

Mária Mörtl [1], Mariem Damak [1,2], Miklós Gulyás [1], Zsolt István Varga [1,2], György Fekete [1,2], Tamás Kurusta [3], Ádám Rácz [3], András Székács [1,*] and László Aleksza [1,2,*]

1 Institute of Environmental Sciences, Hungarian University of Agriculture and Life Sciences, Páter Károly u. 1, H-2100 Gödöllő, Hungary; mortl.maria@uni-mate.hu (M.M.); damak.mariem@phd.uni-mate.hu (M.D.); gulyas.miklos@uni-mate.hu (M.G.); varga.zsolt.istvan@uni-mate.hu (Z.I.V.); fekete.gyorgy@uni-mate.hu (G.F.)
2 Profikomp Environmental Technologies Inc., Kühne Ede u. 7, H-2100 Gödöllő, Hungary
3 Institute of Raw Material Preparation and Environmental Processing, University of Miskolc, Egyetem út 1, H-3515 Miskolc, Hungary; tamas.kurusta@uni-miskolc.hu (T.K.); adam.racz@uni-miskolc.hu (Á.R.)
* Correspondence: szekacs.andras@uni-mate.hu (A.S.); aleksza.laszlo@uni-mate.hu (L.A.); Tel.: +36-28-522-000 (ext. 1721) (A.S.); +36-28-512-490 (L.A.)

Abstract: In recent years, the environmental impacts of plastic production and consumption have become increasingly significant, particularly due to their petroleum-based origins and the substantial waste management challenges they pose. Currently, global plastic waste production has reached 413.8 million metric tons across 192 countries, contributing notably to greenhouse gas emissions. Bioplastics have emerged as eco-friendly alternatives, with bioplastic carrier bags composed of 20% starch, 10% additives, and 70% polybutylene adipate terephthalate (PBAT) being the focus of this research. This study aimed to evaluate the biodegradation of these bioplastic bags under industrial composting conditions, addressing the gap in the existing literature that often lacks real-world applicability. A large-scale composting experiment was conducted using 37.5 tons of manure/wood and 50 tons of biopolymer bags over 12 weeks. Results showed that compost temperatures peaked at 70 °C and remained above 50 °C, pH levels stabilized at 8.16, and electrical conductivity was recorded at 1251 μs cm^{-1}. Significant changes were observed in key metrics, such as the carbon-to-nitrogen ratio and organic matter content. Disintegration tests revealed that 95% of the bags disintegrated by the 12th week, though ecotoxicity tests indicated varying germination inhibition rates. Advanced analytical methods (Fourier transform infrared spectroscopy, gas chromatography coupled with mass spectrometry) highlighted morphological and chemical transformations in the bags. This research enhances understanding of bioplastic degradation in real-world composting environments and suggests potential improvements to existing standards, promoting sustainable waste management solutions.

Keywords: PBAT; biopolymer; composting; biodegradation; FTIR; GC-MS

1. Introduction

Plastics consist mainly of synthetic polymer compounds derived mostly from petrochemical sources. These compounds are characterized by their high molecular mass, inert, hydrophobic, and long-chain polymers, and their resistance to biodegradation in nature [1]. Inadequate disposal of plastic waste has been linked with various negative environmental impacts. Global plastic waste production reached 413.8 million metric tons across 192 countries in 2023 [2]. In 2022, an estimated 21 million tons of macroplastics leaked into the environment globally, over 30% more than during the previous decade, and plastics are estimated to contribute 3.8% of total global greenhouse gas emissions [3]. Consequently, plastic pollution has now been identified in all major ocean basins, beaches, rivers, lakes, terrestrial habitats, and even isolated regions such as the Antarctic and Arctic regions [4]. Furthermore, plastics are a significant contributor to greenhouse gas (GHG) emissions

throughout their life-cycle. According to the Organization for Economic Co-operation and Development [4], 1.8 billion tons of GHGs were emitted in 2019 in relation to the plastic industry, where 90% of these emissions resulted from the production of plastics and their conversion from fossil fuels. These emissions are also linked with detrimental health impacts on humans, such as by the emission of particulate matter [5]. Additionally, GHGs are emitted at the end-of-life of plastics during waste management or incineration and littering. The latter is well known to pose a significant threat not only to the marine environment but also to the safety of animals and humans through the bioaccumulation and toxicity of microplastics in the food chain [6–11].

Bioplastics or biopolymers, advertised as environmentally friendly and sustainable materials, have emerged as a viable alternative to conventional plastics. According to European Bioplastics, biopolymers may be divided into three types depending on their resources and biodegradability: (1) bio-based and biodegradable, (2) bio-based but not biodegradable, and (3) biodegradable plastics that do not derive from renewable energy sources [12]. Bio-based polymers derived from renewable resources are created by extraction from biomass molecules (such as sugars found in plants) and then polymerized to create either a direct substitute for an existing plastic type, such as polyethylene, or a unique polymer, such as polyhydroxyalkanoate [13]. Non-synthetic natural polymers are typically starch, natural rubber, and proteins that may be extracted from biomass of different plants, including rice, corn, tapioca, potatoes, soybeans, wood cellulose, wheat fiber, and bagasse [14,15].

On the other hand, synthetic fossil-based biodegradable polymers are derived essentially from crude oil but also from coal and natural gas. Degradability is obtained in these polymers by including unstable linkages (ether, amide, or ester bonds) that are vulnerable to hydrolytic attack [16]. The most important and widely produced fossil-based biodegradable biopolymers are polybutylene adipate terephthalate (PBAT) and polycaprolactone [12].

The biodegradation of biopolymers is a process of mineralization of organic materials by various types of microorganisms (e.g., bacteria, fungi, and archaea). The process involves three main steps [17], starting with the deterioration of the matrix, which consists of the alteration of the mechanical, chemical, and physical characteristics of a polymer. In nature, this process is induced by heat activation, hydrolysis, biological activity, oxidation, photolysis, or radiolysis [18]. The second step of the biodegradation process is the bio-fragmentation step, during which the conversion of polymers by the action of microorganisms to oligomers and monomers occurs. Bio-fragmentation occurs extracellularly, where extracellular enzymes' active sites form complexes with the substrate and cleave a portion of it off [19]. Finally, the microbial assimilation of the products of the second step takes place. At this point, the microorganisms are supplied with carbon, energy, and nutritional supplies and convert the biopolymers to water, CO_2, and biomass.

The biodegradation of biopolymers is highly affected by the surrounding environmental factors, as they influence the microbial population and the activity of the individual microorganisms themselves. Humidity, temperature, pH, salinity, the presence or absence of oxygen, and the availability of various nutrients all have a substantial impact on the microbial breakdown of polymers [20]. The process is also affected by the chemical and physical properties of the polymer, including diffusivity, porosity, morphology, cross-linking, purity, chemical reactivity, mechanical strength, thermal tolerance, and electromagnetic radiation resistance [18].

At their end-of-life, biopolymers provide a variety of waste management methods that can generate added value, contributing thereafter to the objectives of a circular economy. Due to their biodegradability, biopolymers such as PBAT can be recycled in some cases [21] or treated as organic waste and can be composted or discarded in anaerobic digestion factories for biogas production [22].

Composting is a controlled, aerobic process that converts organic materials into a nutrient-rich soil amendment and plant fertilizer. The process is mediated by a consortium of microorganisms whose metabolism depends on a variety of environmental conditions

mentioned above, with high importance on the carbon-to-nitrogen (C/N) ratio and the characteristics of the input raw materials [23]. The process can be separated into three main phases: the mesophilic, the thermophilic, and the maturation phases. The mesophilic phase is carried out by mesophilic microorganisms that thrive under moderate temperatures (up to 40 °C) where simple organic compounds such as sugars are mineralized. Exothermic reactions are generated by metabolic activity, raising the composting temperature to about 65 °C [24]. At this stage, the mesophilic microorganisms are exhausted, and the composting pile is colonized by thermophilic microorganisms, marking the second phase of the composting process. The thermophilic stage is the most important phase in composting as elevated temperatures in the compost pile kill pathogenic organisms and weed seeds; moreover, microbes break down complex molecules such as proteins, fats, and complex carbohydrates, including cellulose and hemicellulose [25]. As the energy supply in the compost pile depletes, mesophilic microorganisms recolonize the composting pile in the last phase as the temperature falls below 40 °C. In this stage, the remaining carbohydrates, cellulose, and hemicellulose, are broken down by the microorganisms, and the precursors of humic substances are formed.

Biopolymers can be biodegraded in composting facilities combined with other types of organic waste only if they are certified according to the standards listed below in Table 1.

Table 1. Overview of the published standards for biodegradation of biopolymers under aerobic biological treatment environments.

Standard	Name	Geographical Validity
EN 13432:2001 [26]	Packaging. Requirements for packaging recoverable through composting and biodegradation. Test scheme and evaluation criteria for the final acceptance of packaging	European Union
EN 17033:2018 [27]	Plastics—Biodegradable mulch films for use in agriculture and horticulture—Requirements and test methods	
EN 14995:2006 [28]	Plastics—Evaluation of compostability—Test scheme and specifications	
EN 14045:2003 [29]	Packaging—Evaluation of the disintegration of packaging materials in practical oriented tests under defined composting conditions	
EN 14806:2005 [30]	Packaging—Preliminary evaluation of the disintegration of packaging materials under simulated composting conditions in a laboratory-scale test	
ISO 17088:2021 [31]	Plastics—Organic recycling—Specifications for compostable plastics	Worldwide
ISO 16929:2021 [32]	Plastics—Determination of the degree of disintegration of plastic materials under defined composting conditions in a pilot-scale test	
ISO 18606:2013 [33]	Packaging and the environment—Organic recycling	
ISO 20200:2023 [34]	Plastics—Determination of the degree of disintegration of plastic materials under simulated composting conditions in a laboratory-scale test	
ISO 17556:2019 [35]	Plastics—Determination of the ultimate aerobic biodegradability in soil by measuring the oxygen demand in a respirometer or the amount of carbon dioxide evolved	
ASTM D6400:2022 [36]	Standard Specification for Labeling of Plastics Designed to be Aerobically Composted in Municipal or Industrial Facilities	USA
ASTM D6868:2021 [37]	Standard Specification for Labeling of End Items that Incorporate Plastics and Polymers as Coatings or Additives with Paper and Other Substrates Designed to be Aerobically Composted in Municipal or Industrial Facilities	
AS 43736:2006 [38]	Biodegradable Plastic—Biodegradable Plastics Suitable for Composting and other Microbial Treatment	Australia

The European standard EN 13432 [26] provides specific requirements for determining the compostability and anaerobic treatability of packaging and packaging materials by addressing four characteristics:

(1) Biodegradability: Referring namely to the conversion of organic materials to CO_2 by microorganisms. The standard includes a required biodegradation criterion of at least 90% that must be attained in fewer than 6 months (laboratory test method EN 14046 [39]);
(2) Disintegration during biological treatment: Tested through the fragmentation and loss of visibility in the final compost, which is determined in a pilot composting test (EN 14045 [29]) in which specimens of the tested bioplastic material are composed with biowaste for 12 weeks. Following this period, the mass of the test material residues must be less than 10% of the original mass;
(3) Effect on the biological treatment process: There should be no negative effects of the packaging materials on the composting process;
(4) Effect on the quality of the resulting compost: The increased packing material should not degrade the final product's quality. The standard requires that this be verified using ecotoxicity tests: This entails conducting a study to determine whether composted packaging has a harmful effect on plant germination and biomass output. Moreover, the standard sets limits for volatile matter, heavy metals, and fluorine contents of the final compost.

Composting bioplastics has received much attention recently from scientific studies, particularly with regard to bioplastic carrier bags. Nevertheless, by filling in the knowledge gaps and offering a more thorough grasp of the topic, this work adds fresh perspectives to this field of study. Our research uniquely emphasizes both degradation and disintegration, specifically within the context of industrial-scale composting conditions. While numerous studies have explored the biodegradation of bioplastics, our research topic focuses on the intricacies of both degradation and disintegration processes under the particular conditions seen in large-scale composting operations. The complexities and variables inherent in industrial-scale composting, including elevated temperatures, varying aeration levels, organic matter (OM) content, moisture content (MC), ratio of (C/N) in the organic materials, the dynamic interplay of heterogenous organic waste materials, and the diverse microbial interactions, are often overlooked in laboratory-scale studies. By positioning our research within this industrial framework, our primary objective is to generate findings that more accurately mirror real-world scenarios. This aims to enhance the applicability and relevance of our results in the management of bioplastic waste.

Our paper assesses the biodegradation and disintegration of bioplastic in a significant amount of carrier bag waste, marking a departure from the prevailing focus on small quantities in the existing literature. The prevalent limitation of composting smaller quantities of plastics poses a potential impediment to the effective integration of bioplastics into the compost matrix. By concentrating on larger quantities in our research, we aim to enhance the understanding of bioplastic behavior within more realistic composting environments. Furthermore, our study examines, in agreement with the corresponding standards, fragments both greater and less than 2 mm in size, contributing to a comprehensive analysis of bioplastic dynamics during the composting process.

Moreover, one of the main objectives of this paper is to investigate the impact of various abiotic process conditions on the components of PBAT-based bioplastic film through the use of analytical techniques Fourier transform infrared spectroscopy (FTIR), gas chromatography coupled with mass spectrometry (GC-MS), scanning electron microscopy (SEM), and ecotoxicity tests facilitating a comprehensive analysis of the composting process and guaranteeing a thorough exploration of potential environmental implications, particularly in relation to microplastics. FTIR and GC-MS methods were employed to investigate the composting progression of PBAT-based bioplastic bags. The goal was to identify decomposition products and monitor their concentration variations throughout the composting process. This also involves the detection of possible marker component(s), which are suitable to indicate that the decomposition went into completeness.

2. Materials and Methods

2.1. Tested Material

The test materials used in this study were PBAT-based bioplastic carrier bags available in an international supermarket network in Hungary and certified as compostable under industrial composting conditions according to OK compost® Vincotte by TÜV Austria (Brunn am Gebirge, Austria). Based on the work of Borchani et al. [40], the given PBAT-based biopolymer consists of 20% starch, 10% additives, and 70% PBAT. The studied biopolymer films' width was 143 μm with haze optical properties of 94% according to ASTM D1003 [41]. The characteristics of the studied biopolymer according to the available material datasheet [42] are presented in Table 2.

Table 2. Characteristics of the studied biopolymer.

Characteristic	Value	Test Standard
Melting Temperature	110 °C	ASTM D3418 [43]
Density	1270 kg m^{-3}	ASTM D792 [44]
Tensile Modulus	350 MPa	ASTM D882 [45]
Elongation at Break	230%	ASTM D882 [45]

2.2. Composting Experiment and Sampling

The composting experiment was carried out in an aerated static pile (ASP) located near Orosháza (Békés County, Hungary). The membrane-covered, side-walled ASP system was an industrial-scale composting unit with dimensions of 25 m in length and 8 m in width (Figure 1). The initial raw materials mixture was composed of approximately 37.5 tons of cattle manure (used as a nitrogen source to adjust the C/N ratio) and shredded wood (used as a bulking agent) and 50 tons of biopolymer carrier bags. Fresh cattle manure with minor straw content was obtained from a local farm in Orosháza city with a production capacity of 5000 tons per year. At the beginning of the experiment, the biopolymer carrier bags were shredded into 10 mm particles and mixed with the wood and manure raw materials. The composting technology used during this study was an aerated static pile composting system with a semipermeable membrane cover, the dimensions of which were 8 m by 4 m. The active aeration system supplied the oxygen needed for the activity of microorganisms during the biodegradation process and was delivered by perforated pipes at the bottom of the compost pile. The aeration system was continuously controlled by the feedback information about the temperature and oxygen levels of the compost pile. The membrane cover allows gas exchange while it retains unpleasant odors, moisture, and heat. The temperature variation during the 12-week composting experiment was measured with a SCADA temperature probe with a length of 1000 mm and equipped with a pt100 electrode (Profikomp Inc., Gödöllő, Hungary).

(a) (b)

Figure 1. *Cont.*

(c)

Figure 1. Composting of polybutylene adipate terephthalate-based bioplastic carrier bags, composed of 20% starch, 10% additives, and 70% polybutylene adipate terephthalate at the industrial scale (50 tons), in a membrane-covered side-walled aerated static pile (ASP) system covered with expanded polytetrafluoroethylene membrane cover ProfiCover®, Profikomp Environmental Technologies Inc., Gödöllő, Hungary), located near Orosháza, Békés County, Hungary. (**a**,**b**) Composting setup before the experiment. (**c**) ASP during composting.

As for the large amount (50 tons) of the waste biopolymer material and its ratio to the cattle manure applied (1.33), the objective of our study was to assess the biodegradation and disintegration of bioplastic in a substantial, industrial-scale amount of PBAT waste material marking a departure from the prevailing focus on small quantities in the existing scientific literature. The prevalent restraint of composting smaller quantities of plastics poses a potential impediment to the effective integration of bioplastics into the compost matrix. The limitations of the usual practice of composting biopolymers in small batches or including a minor ratio (in most cases less than 1%) in the composting mixture contribute to the misunderstanding of waste treatment principles, as the dilution, which deliberately changes the characteristics of waste without improving its environmental performance, should not be allowed. In our experiment, we explored the potential upper limit of the economical and professional optimum of composting biopolymers as regular composting facilities usually do not accept biopolymer waste due to the required permissions, low economic profitability, and impact on regular operation, putting this type of waste utilization at risk. Moreover, we attempted to apply as little manure as possible but still remain in the workable range of C/N ratios (25 to 35) [46–48]. By concentrating on larger quantities in our research, we aim to enhance the understanding of bioplastic behavior within more realistic composting environments.

Sampling was carried out in triplicates on a biweekly basis from the beginning of the experiment until the 12th week. Samples were collected from 1.0 ± 0.2 m depths inside the pile in the middle section to avoid the surface effect. Each time, sample volumes of 15–30 L per point were collected and homogenized to obtain an average sample of 15–20 L by sample reduction. The composite samples from replicates were labeled with sequential even numbers from 0 to 12, representing the number of the composting week. The finished compost was sieved through a 10 mm sieve and placed for maturation for 1 year. Sieving was carried out carefully by hand to avoid, as much as possible, systematic errors due to induced mechanical fragmentation. Physical breakdown was assessed by optical microscopy (see Section 3.6.1). A composite sample was taken from the 1-year matured compost and named M1.

As the original PBAT-based waste was shredded before substrate mixing for composting, the process resulted in 10 mm wide plastic strips with varying lengths. Due to the nature of the PBAT-starch-based waste material, in the first few weeks, the thermophilic conditions made the substance soften up and curl/aggregate with the bulking agent and manure constituents. Subsequently, it hardened up and became crumbling, resulting in disintegration, as stated later (see Sections 3.3 and 3.4). Because of the initial shredding and high (yet not full) level of disintegration determined at the end of the 12th week, the oversize fraction did not contain a significant amount of polymer after sieving (only the bulking agent) and was excluded from analysis due to the fact that is was aimed to be used in the next batch of composting mixture.

2.3. Physicochemical Properties of the Input Material and Monitoring During the Composting Process

Various composting parameters of the input mixture were measured at the beginning, throughout, and at the end of the composting process. Moreover, compost quality parameters were measured after the 1-year maturation period. The values provided in the standard EN 14045:2003 [29] were used as references. The parameters measured and the methods used for the analysis are summarized in Table 3.

Table 3. Measured parameters during the study.

Measured Parameter	Test Method
Carbon-to-nitrogen ratio of the input material (C/N)	Elemental analysis to measure the carbon and nitrogen contents of the initial mixture MSZ EN ISO 16948:2015 [49], MSZ EN ISO 16994:2017 [50]
Organic matter (OM)	Loss on ignition (MSZ EN 15935:2012) [51]
Moisture content (MC)	EN 16086-1:2012 [52]
pH	Potentiometric determination of H^+ ion concentration, MSZ EN 13037:2012 [53]
Electrical conductivity	ISO 7888:1985 [54]
Salt content	ISO 8502-6:2020 [55]
Bulk density	ISO 8502-9:2020 [56]
Respiration intensity	Malinska, K. (2016) [57]

2.4. Disintegration Test

To evaluate the disintegration behavior of the studied biopolymers throughout the composting process, the standard ISO 16929 [32] was used as a reference, and the sieving analysis was carried out on the first day, 2nd, 4th, 6th, 8th, 10th, and 12th week of the composting process. Each sample was sieved through a 10 mm sieve and examined for big lumps of compost containing fragments of the test material, which were broken up into crumbly particles. The sieved material was further separated through a 2 mm sieve. Test material particles were then collected from the obtained 2 mm to 10 mm fractions, cleaned, dried at 40 °C until constant mass was reached, and finally weighed.

The disintegration rate of each sample throughout the composting process was evaluated based on the respective total dry solids by comparing the mass of the selected test materials from the >2 mm sieved fractions to the mass of the initial test material input (Equation (1)) [32]:

$$D_i = \frac{m_1 - m_2}{m_1} \times 100\% \quad (1)$$

where Di is the degree of disintegration of the test material (in %), m_1 is the mass of total dry solids in the test material input (in g), and m_2 is the mass of the total dry solids in the retrieved test material (in g).

2.5. Morphological Changes

2.5.1. Sample Preparation

The disintegration test conducted on a biweekly basis generated two biopolymer compost fractions through the 2 mm sieve by the end of each test: a fraction > 2 mm and a fraction < 2 mm. Plastic fragments from the >2 mm biopolymer compost fraction were recovered, dried, and carefully toothbrushed to remove all deposits formed on their surfaces. The <2 mm fraction was separated into 4 different fractions by sieving it through 1 mm, 500 μm, and 106 μm standard sieves. Microplastics over 500 μm were collected with tweezers as it was possible to perform the process with the naked eye. They were then dried and very carefully toothbrushed.

2.5.2. Optical Microscopy

As it was not possible to investigate with a naked eye the compost matrix combination under 500 μm, it was examined with an AxioCam MRc 5 optical microscope (Carl Zeiss Microimaging Inc., Göttingen, Germany). Pictures were taken with the built-in digital camera (Carl Zeiss Microimaging Inc., Göttingen, Germany).

2.5.3. Scanning Electron Microscopy (SEM)

Examining the composting stages of PBAT-based bioplastic carrier bags on a microscopic level is crucial for comprehending the degradation process and assessing disintegration efficiency. Images of intact and degraded bioplastic fragments from the >2 mm biopolymer compost fraction and the microplastics recovered from the <2 mm fraction (see Section 2.5.1) were collected with HITACHI S-4700 field emission-scanning electron microscope (Hitachi High-Tech Corp., Tokyo, Japan). The biopolymer fragments were metalized with a 10 nm coating of gold before microscopy.

2.6. Fourier Transform Infrared (FTIR) Spectroscopy

To investigate the chemical composition changes of the biopolymer bags throughout the composting process, spectra of the intact and degraded fragments from the >2 mm biopolymer compost fraction and the microplastics recovered from the <2 mm fraction (see Section 2.5.1) were collected in total reflectance mode (ATR) with Jasco FT/IR-4200, equipped with a diamond prism and an ATR Pro Penta device (Jasco Corp.; Easton, Saint Louis, MO, USA).

2.7. Gas Chromatography Coupled with Mass Spectrometry (GC-MS)

Samples taken after 2, 6, and 12 weeks of composting, as well as 1-year matured compost, have been subjected to GC-MS. The comparative analysis between the larger (>2 mm) and smaller (<2 mm) fractions was emphasized to understand the extent and nature of biodegradation and disintegration.

Sample preparation prior to GC-MS measurements started with the extraction of 1 g sample by using 10 mL of acetonitrile. Samples were subjected to ultrasound agitation for 15 min in a Falcon tube. One mL of the extract was then filtered by a syringe filter (0.45 μm), and the solvent was evaporated by nitrogen stream. A total of 50 μL of silylating agent (*N,O*-bis(trimethylsilyl)trifluoroacetamide, BSTFA) was added to the dried residues and derivatized for 30 min at room temperature under moisture-free conditions. Finally, 10 μL of internal standard, containing eicosan (C20, ISTD) at the level of 1 mg mL^{-1}, was added to the reaction mixture and diluted to 1 mL with hexane. After homogenization by vortex for 30 sec, 3 μL of the sample was injected. Adipic acid (AA), 1,4-butanediol (BDO), and phthalic acid (PTA) standard solutions were derivatized using the same procedure to establish the calibration curves. Levels for external calibration were 0.5, 1.0, 2.5, 5.0, and 25 μg mL^{-1}, which are equal to the target component content of 5 to 250 μg g^{-1}. Estimated amounts of the other three components containing either AA or PTA and additional butoxy unit(s) (AA+, AA++, and PTA+) based on the assumption that their detector responses are similar to the corresponding monomer acids (AA and PTA). Validity could not be checked

as the standards were not available for these components, but changes in signal intensities obviously show the trends related to their quantity.

Determination of compounds was performed on a Chromass (YL Instruments, Anyang, Republic of Korea) instrument. The initial temperature was 80 °C, held for 4 min, then increased by 15 °C min^{-1} to 280 °C, held for 3 min. The injection port was kept at 250 °C, and the transfer line at 280 °C. The solvent delay was 5.5 min; thereafter, fragments in the range between 50 and 400 m/z were detected. Components were separated on a TG-5 ms column with the dimensions of 0.25 mm × 0.25 µm × 30 m (ThermoScientific, Waltham, MA, USA).

2.8. Ecotoxicity Tests

To determine the effect of the biopolymer compost on the germination and growth of plants, two plant bioassays were implemented: (1) EN 16086-1 [52]: soil improvers and growing media. Determination of plant response pot growth test with Chinese cabbage (*Brassica napa* ssp. *pekinensis*) and spring barley (*Hordeum vulgare*); and (2) the Hungarian standard MSZ 08-0012-4 79 [58], Germination dynamics white mustard (*Sinapis alba*). The 5-day stage is the first validation point of the test, as over 85% of the seeds germinated as needed. The endpoint is phenological stage-related, depending on the control, meaning a differing end point of the tests, 16 days for spring barley and 37 for Chinese cabbage, respectively.

Germination inhibition (*GeI*) is expressed as a percentage of the average germination in the control samples according to Equation (2) EN 16086 [52]:

$$GeI\ (\%) = \frac{AGR_{contol} - AGR_{sample}}{AGR_{control}} \times 100\% \quad (2)$$

where *AGR* is the average germination rate calculated as the average of the germinated seeds in the different pots after five days from the beginning of the experiment.

3. Results and Discussion

3.1. Carbon-to-Nitrogen Ratio

The C/N ratio, widely accepted as the most prognostic descriptor of the composts for their N mineralization capacity [59], was measured for the two main components: cattle manure with very low straw content (10:1) and the biopolymer sample (504:1). After correction with the weight and moisture content of the materials a C/N ratio of 35.6 ± 5.3 was detected for the mixture, falling into the range of optimum 25–35 C/N ratio of the initial mixtures [46–48]. The carrier bag waste was treated similarly to paper waste when calculating the mixture using a standard operational protocol. The exact aim was to try out the possible limitations of bioplastic to the composting process. The supplier of the carrier bag waste was interested in the possibility of creating a dedicated composting plant for this type of waste.

3.2. Temperature

The temperature was recorded daily throughout the entire composting experiment (13 weeks), and Figure 2 illustrates the temperature variation curve over time. The temperature curve exhibits an initial temperature of 52 °C. Subsequently, the composting process transitioned directly into a thermophilic phase, characterized by a sustained temperature of greater than 50 °C. The temperatures recorded were similar to the ones observed by Itävaara et al. (1997) [60], who also reported temperatures exceeding 50 °C and rising as high as 80 °C. Another study by Kianirad et al. (2009) [61] also noted that higher temperatures, typically exceeding 55 °C, are favorable for the efficient biodegradation of PBAT, and even longer aromatic oligomers can degrade during composting at such elevated temperatures [62]. Secondly, the compost maintained a pH between 6 and 7, and an increased pH of 8.16 was observed by the end of the study. A high pH of 8 was also observed by Itävaara et al. (1997) [60] at the end of the composting experiment. The observed shift to an

alkaline level likely results from ammonia release, a phenomenon tied to the decomposition of nitrogen-rich compounds. The mature compost's electrical conductivity, measured at 1251 μs cm^{-1} at 22.8 °C, and its salt content of 0.6% highlight the significance of salinity in the composting pile. Tester (1990) [63] underscored that elevated salt levels, particularly sodium chloride, can hamper microbial activity, thereby decelerating decomposition. The salt metrics from our compost suggest a balance that does not hinder its potential utility, yet careful monitoring remains essential.

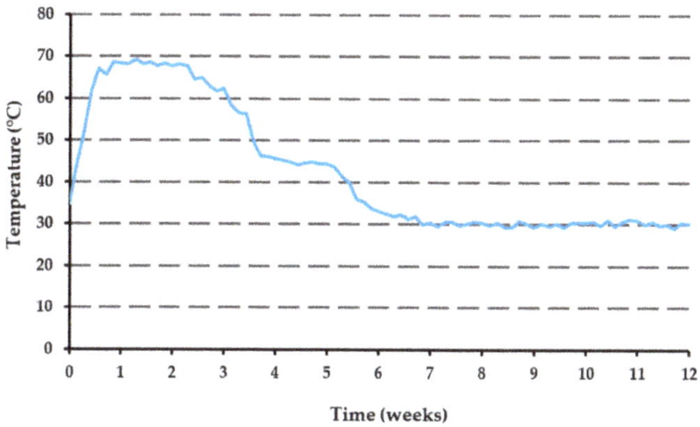

Figure 2. Temperature curve during the 12 weeks of the composting process.

3.3. Respiration Intensity

The oxygen consumption rate was monitored throughout the 12-week composting process to examine the aerobic activity and the rate of decomposition of the biodegradable organic components within the composting material. The oxygen uptake rate is a vital indicator, as it is directly linked to the mass of the degradable substance through a reaction coefficient (k) known, from which the respiration activity, AT4, is derived by first-order reaction kinetics. This coefficient is influenced by various environmental factors, including temperature, MC, biodegradability, fatty acid synthesis, pH, and the initial C/N ratio of the material [64]. Figure 3 illustrates the evolution of the oxygen uptake rate over the composting period. The graph revealed distinct microbial activity patterns during the composting of the bioplastics.

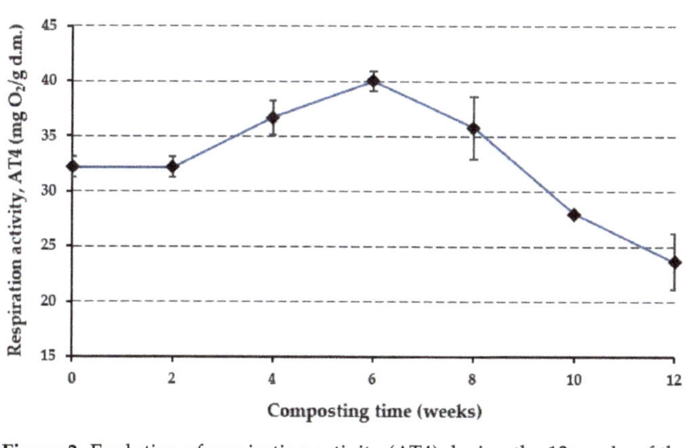

Figure 3. Evolution of respiration activity (AT4) during the 12 weeks of the composting process.

The initial two weeks were characterized by a lag phase, with a constant respiration intensity of approximately 32 mg O_2 g^{-1} dry matter (d.m.). This phase signifies the

acclimatization and adaptation period of the microbial community to the composting environment. After this phase, there was a noticeable increase in microbial activity. The oxygen consumption rate gradually escalated, reaching its peak at about 40 mg O_2 g^{-1} d.m. during the sixth week. After reaching the peak, the respiration intensity began to gradually decrease. This decline continued until the end of the composting process, where the oxygen consumption rate was finally measured at 23 mg O_2 g^{-1} d.m. This reduction in respiration intensity signifies the depletion of easily degradable materials and the stabilization of the compost. The oxygen consumption rate is closely linked to the CO_2 production rate, as microbes utilize oxygen to metabolize organic materials, emitting CO_2 as a waste product. This balance between oxygen consumption and CO_2 production is indicative of the biodegradability of the material [65]. The results are comparable with the findings of [66], where they reported high degradation rates for specific plastic polymers, even surpassing the biodegradation rate of the cellulose standard. This rapid biodegradation, evident in the early stages of their trials, resulted in over 80% biodegradation in just about 20 days for certain plastics [66]. The results are also comparable to the study by Tabasi and Ajji (2015) [67], where the composting of PBAT also exhibited a lag phase followed by a peak in oxygen consumption.

3.4. Organic Matter Mineralization

The initial C/N ratio as the basis of mineralization was slightly higher than the literature recommendations and gradually decreased to 15.9 ± 0.3 during the composting process. The mineralization and degradation rate of organic matter (OM) are pivotal indicators of compost quality and stability. At the commencement of the composting experiment, the OM content was recorded at 77.43% ± 9.47. Throughout the 12-week period, a consistent reduction in this value was observed, culminating in an OM content of 62.43% ± 1.25 by the end of the process. This signifies a relative loss of 17.39% ± 10.72 in OM.

The concluding OM content denotes a high-caliber compost product, signifying proficient decomposition and stabilization of OM. The values for OM content during the experiment are provided in Table 4. The results observed were compatible with the findings by Ruggero et al. (2021) [68], who employed the Walkley–Black method for the measurement of total organic carbon, providing a benchmark for OM content.

Table 4. Organic matter (OM) and moisture content (MC) during the experiment.

Composting Week	OM (%)	MC (%)
0	77.43 ± 9.47	46.08 ± 1.29
2	74.09 ± 6.77	41.46 ± 0.17
4	75.60 ± 1.46	41.15 ± 1.20
6	73.74 ± 8.18	46.42 ± 1.25
8	69.79 ± 1.32	40.81 ± 0.02
10	65.07 ± 3.81	36.97 ± 0.77
12	62.43 ± 1.25	35.90 ± 1.36
M1 [a]	60.04 ± 1.12	5.46 ± 1.21

[a] 1-year matured compost.

The composting process showed a consistent decline in MC over its duration. The initial MC was measured at 46.08% ± 1.29. By the 12th week, this value had decreased to 35.90% ± 1.30. A significant positive correlation between microbial activity and MC was observed, with a correlation coefficient of R^2 = 0.75 ($p < 0.05$). This result emphasizes the direct relationship between moisture and microbial activity, suggesting that the MC plays a pivotal role in the microbial degradation process during composting. Drawing from Ruggero et al. (2020) [69], the biodegradation of the PBAT-based biopolymer is highly influenced by moisture. Under conditions with MC between 55% and 45%, PBAT showed degradation rates of up to 90.4%. When MC drops below optimal values, there is a noticeable slowdown in biological activity, which can also be reflected in the degree of

biodegradation. The obtained values were less than the values mentioned in the standard. The mature compost had a low moisture content of 5.46 ± 1.21 due to the long drying period. The OM and MC values during and at the end of the composting are provided in Table 4.

3.5. Disintegration Test

The degree of disintegration of the PBAT-based biopolymer during the 12-week composting process was calculated using Equation (1), and the values are presented in Figure 4. A substantial increase in the degree of disintegration, particularly noticeable from the fourth week onwards, was observed. By the 12th week, the degree of disintegration reached 95% ± 1.86. A notable bioplastic disintegration of 67.57% ± 10.21 was observed after the second week of the process and reached 78.82 ± 13.49 by the fourth week. Figure 5 illustrates the diminished presence of plastic fragments larger than 2 mm at this stage. Figure 5c represents the disintegrated fragments at the end of the composting experiment. The study by Ruggero et al. (2020) [69] also found that bioplastics underwent disintegration of more than 90% by the end of composting.

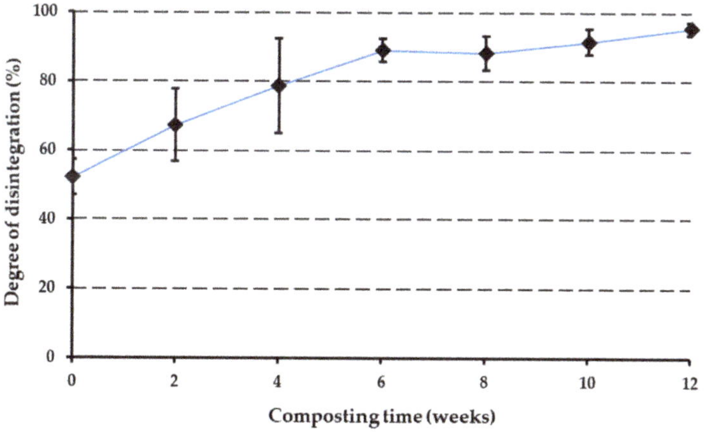

Figure 4. Degree of disintegration of biopolymer during the 12 weeks of the composting process.

(**a**) (**b**) (**c**)

Figure 5. Representative pictures taken during the disintegration of the biopolymer fraction greater than 2 mm (**a**) on the first day, (**b**) on the 6th week, and (**c**) on the 12th week of the composting process.

A visual examination of the sieved compost revealed discernible contaminants, significantly affecting the overall coloration and rendering it a grayish hue (Figure 6). To further investigate, an additional image, Figure 6, was captured after a maturation period of one year, aiming to assess any visible changes in the compost's appearance. Evidently, the grayish aspect persisted, visually resembling microplastic fragments.

Figure 6. Visual compost matrix after 1 year of maturation.

3.6. Morphological Changes

3.6.1. Optical Microscopy

Mechanical fragmentation from the sample preparation process could be distinguished from biological disintegration by examining the edges of particles under a microscope. Microscopic evaluation indicated that the effect of sieving was practically negligible yet an unavoidable consequence of sample preparation. Plastic fragments smaller than 2 mm were difficult to detect and extract from the compost matrix. Figure 7 provides images of various compost matrix fractions captured during the second and twelfth weeks using an optical microscope at different magnifications. Due to the matrix's intricate composition, the microscopic identification of plastic materials was challenging. That is why SEM was used to further detect the plastic fragments and their degradation over time. Identification was primarily based on shape and color attributes, as highlighted by Jung et al. (2021) [70]. As evident in Figure 7a–d, the compost matrix contained particles in shades of white, cream-white, green, and red/pink. These particles often displayed fragmented or flake-like appearances with uneven edges. Additionally, transparent particles, potentially microplastics, were present but posed identification challenges.

Figure 7. *Cont.*

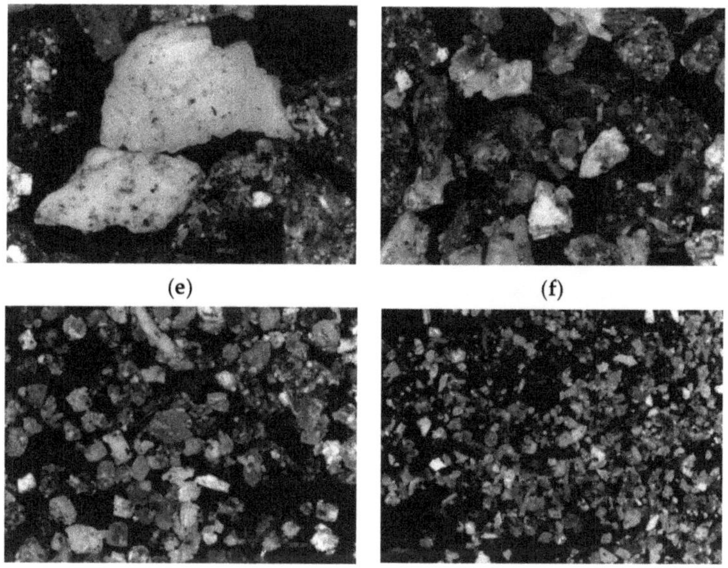

(e) (f) (g) (h)

Figure 7. (**a–d**) Images of fragments retained in the compost matrix after the 2nd week of composting; (**e–h**) images of fragments retained in the compost matrix after the 12th week of composting.

The potential presence of unknown additives in the bioplastic matrix further complicated the identification process, emphasizing the limitations of optical microscopy for recognizing biodegradable microplastic fragments. Figure 7e–h represent the compost matrix after 12 weeks. The finished compost appeared to contain fewer large fragments (>500 μm) but increased smaller fragments in fractions < 250 μm. Additionally, the images revealed the formation of a more complex matrix of OM and mineral constituents. This matrix was characterized by the absorption of biodegradable microplastics into other organic substances originating from the manure. As dispersity increases (decreases in particle radii), the active surface area typically increases, potentially leading to enhanced adsorption capacity for nanoplastics, as suggested by Fojt et al. (2020, 2022) [71,72]. These interactions could hinder the microbial degradation of bioplastics. Thus, understanding the formation of a complex matrix during composting and the dynamics of microplastic adsorption is crucial for effectively mitigating microplastic pollution from biodegradable polymers.

Moreover, the reliance on attributes such as shape and color for identification well corresponds to findings presented by Ruggero et al. (2020) [69], where visual inspections indicated marked changes in the bioplastic materials just after 5 days of exposure to organic waste. In their study, the bioplastic pieces showed significant OM deposition, emphasizing the interactions between bioplastics and surrounding OM, which may facilitate microbial exchanges.

3.6.2. Scanning Electron Microscopy (SEM)

The SEM micrographs presented in Figure 8, captured at varying magnifications, display the PBAT-based bioplastic bag's intricate microstructure. Specifically, these images highlight the presence of spherical spots of a few hundred nanometers. These spherical inclusions are identified as the starch fraction of the blend mix, dispersed within a continuous PBAT matrix [73]. The size of spherical spots depends on the exact composition of the biopolymer, and the presence of additives (e.g., compatibilizers) affects the microstructure as well as the degradation rate [74].

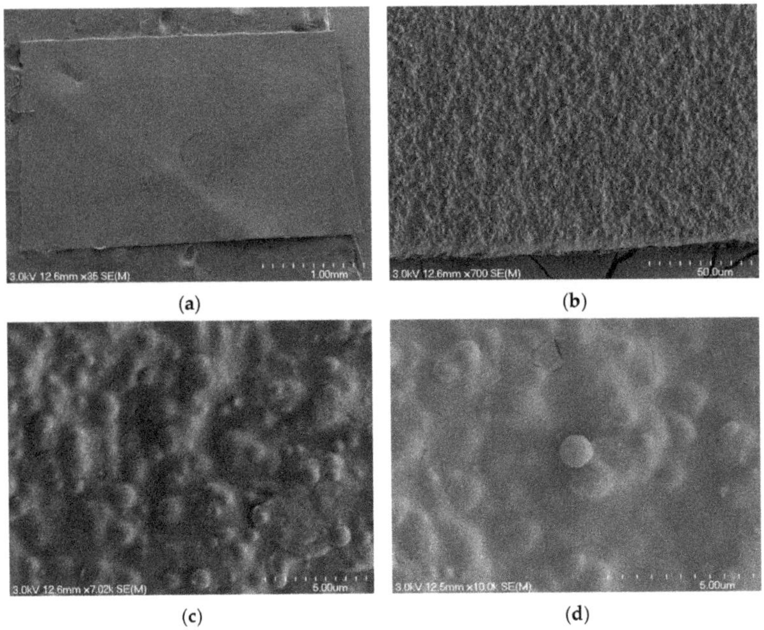

Figure 8. Scanning electron microscope micrographs of a sample material captured at varying magnifications. (**a**) 35× magnification offering a broad overview of the surface. (**b**) 700× magnification showcasing a unique fibrous texture. (**c**) 7.02 k magnification detailing intricate surface nodules and irregularities. (**d**) 10.0 k magnification showing granular formations and the intricate nature of the surface.

Figure 9 shows the morphological alterations observed in carrier bag fragments, sourced from both >2 mm and <2 mm compost fractions, at the 6th and 12th weeks of the composting process. By the sixth week, the onset of surface disruptions, evidenced by cracks, becomes noticeable on the formerly smooth biopolymer bag's surface (Figure 9a,c). This degradation escalates by the 12th week, manifesting as heightened surface irregularities and more pronounced cracking (Figure 9b,d). As the biodegradation process continues, the microstructure transforms into a 3D porous network. Notably, the characteristic starch-based spherical spots diminish and are replaced by nanoscale voids, typically in the magnitude of hundreds of nanometers. This morphological evolution suggests that the starch component of the blend underwent substantial biodegradation during the composting process. The increased surface abrasiveness, coupled with the emergence of fissures and voids, potentially augments sorption sites. This could enhance microplastics' propensity to adsorb specific contaminants or organic molecules [75].

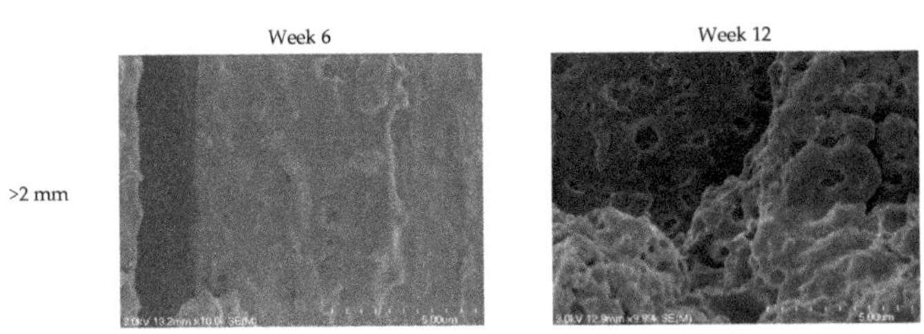

Figure 9. *Cont.*

<2 mm

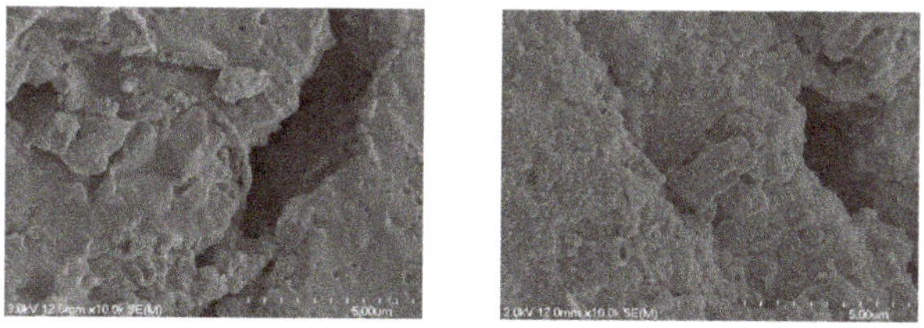

(c) (d)

Figure 9. Scanning electron microscope micrographs depicting morphological changes in carrier bag fragments over the course of composting. (**a**) Week 6 fragment from the >2 mm compost fraction, revealing initial surface disruptions and cracks. (**b**) Week 12 fragment from the >2 mm compost fraction, highlighting intensified degradation with pronounced irregularities. (**c**) Week 6 fragment from the <2 mm compost fraction, showcasing initial signs of wear and surface cracks. (**d**) Week 12 fragment from the <2 mm compost fraction, exhibiting marked surface degradation and increased cracking patterns.

The SEM micrographs of the PBAT-based bioplastic material in the current study, which depicted a heterogeneous microstructure with the presence of circular spots of a few hundred nanometers, align with the observations made by Ruggero et al. (2020) [69]. These spots, attributed to starch constituents, were dispersed within a 3D polymeric matrix, believed to be primarily comprised of PBAT. Such a microstructural depiction resonates with the portrayal of a PBAT-based bioplastic by Ruggero et al. (2020) [69], wherein the circular starch spots underwent progressive disappearance during the composting process. This was accompanied by the emergence of nanoscale holes at the sites of these starch granules, suggesting the composting-induced degradation of starch. Furthermore, Ruggero et al. (2021) [76] and Myalenko and Fedorova (2023) [77] embarked on a comprehensive exploration of bioplastic degradation during composting. Their SEM evaluations corroborate the morphological transitions observed in the current study, reflecting the dynamic microstructural alterations that bioplastics undergo throughout composting. This includes changes in surface roughness, the emergence of cracks, and the development of 3D porous networks.

3.7. Fourier Transform Infrared (FTIR) Spectroscopy

To investigate the chemical composition changes of the biopolymer material during composting and the impact of shredding on their molecular structure, raw PBAT-based bioplastic bag (Figure 10a), as well as fragments from both the >2 mm (Figure 10b) and the <2 mm fractions (Figure 10c), were collected and subjected to FTIR spectrometry analysis in total reflectance mode. The collected fragments were analyzed on a weekly basis, spanning from week 0 to week 12 of the composting process. Figure 10a shows the FTIR spectra obtained from the raw biopolymer bag compared to week 12 of composting, and spectra of the biopolymer bag sample for larger (>2 mm) and smaller (<2 mm) fragments from week 0 to week 12 are presented in Figures 10b and 10c, respectively.

(a)

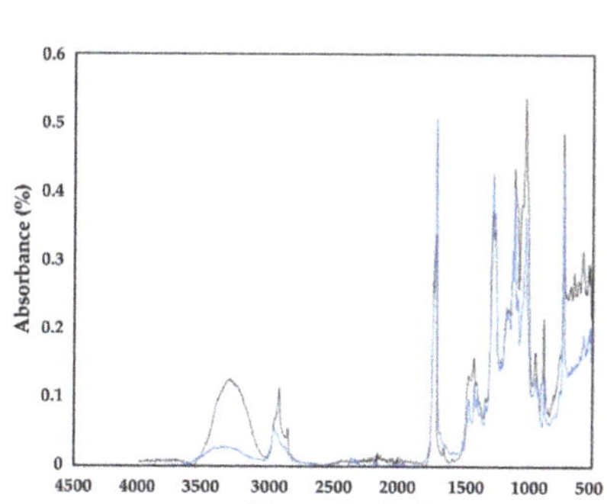

Figure 10. *Cont.*

(b)

(c)

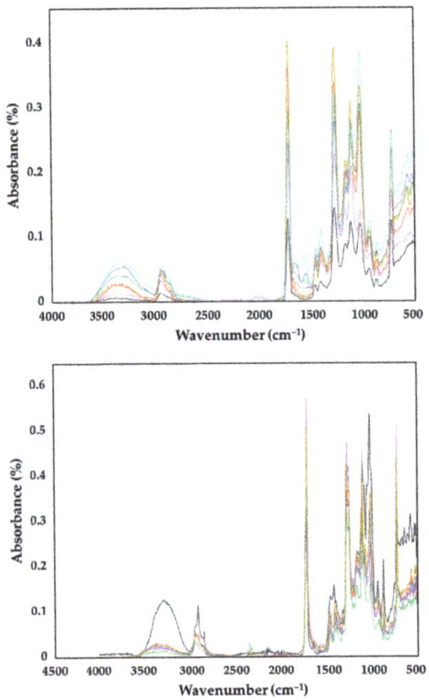

Figure 10. Fourier transform infrared (FTIR) spectra of the polybutylene adipate terephthalate-based bioplastic bag material. (**a**) Comparing the raw sample (gray line) to the sample after 12 weeks (blue line) of composting. (**b**) The bioplastic bag fragments (>2 mm) comparing the raw sample fraction (gray line) to sample fractions after 2 weeks (red line), 4 weeks (blue line), 6 weeks (green line), 8 weeks (purple line), 10 weeks (yellow line), and 12 weeks (turquois line) of composting. (**c**) The PBAT-based bioplastic carrier bag fragments (>2 mm) comparing the raw sample fraction (gray line) to sample fractions after 4 weeks (red line), 6 weeks (blue line), 8 weeks (green line), 10 weeks (purple line), and 12 weeks (yellow line) of composting.

FTIR analysis shows structural changes and biodegradation pathways; however, FTIR spectra of decomposition products (AA, PTA, and BDO) overlap with that of the starting material. In addition, starch has, unfortunately, no characteristic peak out of the crowded fingerprint region. The peaks recorded for the composted samples are in accordance with the literature data reported for PBAT (Table 5). Peaks in the region between 3500 and 3000 cm^{-1} (O-H stretching) indicate the presence of starch fraction in the bioplastic bag [78]. The exact frequency is influenced by the adsorbed MC and associated hydrogen bonds. Asymmetric and symmetric C-H stretching appear at 2957 and 2874 cm^{-1}, respectively. Strong signals observed at 1720 cm^{-1} and 1274 cm^{-1} represent the carbonyl (C=O) and C–O groups in the ester linkages [40]. According to Ruggero et al. (2020) [69], the shifts in the peak at 1717 cm^{-1} are potentially indicative of degradation within the PBAT fraction of the biopolymer and are potentially related to initial hydrolytic actions or enzymatic activities targeting the ester linkages within PBAT, thereby initiating its degradation. Peaks between 1500 and 1300 cm^{-1} are representative of the aromatic structures within PBAT, but weak deformation vibrations of the CH_2 group may also contribute to the signal intensities in this region. Peaks at 1110 and 1025 cm^{-1} correspond to C–O stretches in secondary and primary alcohol moieties, but also, in this region, the in-plane C–H bending of the aromatic rings appears. They exhibit alterations in relative intensities as well as disappearances of shoulders that indicate a gradual degradation or structural modification of the starch fraction within the biopolymer matrix. Signals at 940 cm^{-1} correspond probably to the O–H bend, whereas peaks at 875 cm^{-1} are related to the C–H deformation vibration in

para-substituted phenyl rings. Similar bands are detected for PTA, and there is also a signal at 934 cm^{-1} in the FTIR spectrum of AA.

Table 5. Summary of wavenumbers and their corresponding functional groups.

Peak Wavenumbers in Current Study (cm^{-1})	Vibration	Peak Wavenumbers from Scientific Literature (cm^{-1})	Functional Group Assigned	Material	Reference
725	Asym. deformation	726	$-CH_2-$	PBAT	[40]
1025	Deformation	1016	Phenyl ring	PBAT	[75]
1110	Stretching (alcohols)	1104	C–O	PBAT	[75]
1274	Stretching (ester)	1268	C-O in the ester	PBAT	[40,79]
1410	CH_2CO sym. deform.	1409	$C(-H)_2$	PBAT	[73]
1450	Stretching	1456	Phenylene group	PBAT	[73]
-	-	1505	Benzene ring	PBAT	[75]
1720	Stretching	1710	C=O	PBAT	[40,75]
1270 (shoulder)	Stretching	1270	C–O	Starch	[73]
-	-	1445–1225	C–H	Starch	[80]
3300	Stretching	3900–3300	O–H	Starch	[78]
2874, 2957	Sym./asym. stretching	2920	C–H	Starch	[78]
-	-	1250–900	C–O	Starch	[78]
1150	Shoulder	1164	CH_2OH	Starch	[73]
1085	Shoulder	1081	C–O	Starch	[73]

The comparison between the raw material and the sample at week 12 (Figure 10a) shows slight shifts and changes in intensity, which implies possible ester bond hydrolysis related to bioplastic biodegradation. The spectra of the PBAT-based biopolymer matrix indicate changes in the region of 1700–1750 cm^{-1} attributed to carbonyl (C=O) stretching vibrations. Similarly, the changes in the aliphatic chains of the polymer are observed at peaks around 2957 cm^{-1} and 2874 cm^{-1} (CH_2 stretching). The observed differences in these peak intensities between the raw material and the 12-week sample may indicate potential disruptions of the aliphatic backbone, which could be instigated by abiotic factors or microbial activity during the composting process. In addition, the relative intensity of O–H stretching vibration compared to the initial spectrum substantially decreased as well, and peaks at 1110 and 1025 cm^{-1} were lost from their intensities as starch decomposed.

It is observed that the spectrum for fraction >2 mm (Figure 10a) after 4 weeks of degradation is almost identical to the spectrum recorded at 2 weeks of degradation. However, certain changes emerge during the sixth week of the composting process, with a slight decrease in the intensity of the peaks at 1104 cm^{-1} attributed to C–O vibration and at 1268 cm^{-1} attributed to C–O in the ester linkage. This decrease indicates the start of degradation in the fraction above 2 mm, which means that the polymer's degradation began late at the thermophilic phase.

The FTIR spectra of fragments <2 mm (Figure 10c) show more overall degradation, but the persistence of certain peaks shows that microplastics are still present in the compost matrix. Table 5 presents the measured and reported wavenumbers and the corresponding functional groups discussed in the literature.

Nevertheless, signal intensities in ATR are influenced not only by the absorptivity of functional groups and the concentration of molecules, but other factors (e.g., number of reflections) have an impact. The penetration depth in IR absorbance measurements is only some microns under the material surface; therefore, inhomogeneities of the surface can result in high variability. In addition, the penetration depth is higher for low wavenumbers, which results in differences in relative intensities compared to the transmission spectra. Changes in the 3200–3500 cm^{-1} spectral region, associated with hydroxyl (OH) stretching and water adsorption, might reflect variations in the hydrophilic nature of the material or its interaction dynamics with water molecules during the composting, both of which could be influenced by the degradation processes. The disappearance of peaks or shoulders attributed to starch implies that the fraction has degraded by the end of the composting process. Based on the changes in FTIR, the end of degradation and decomposition rate cannot be determined. As the functional groups in possible metabolites are similar, signals overlap

in the FTIR spectrum of the polymer. Moreover, reference materials, as well as detailed knowledge of the decomposition routes, are lacking. All these prevent the quantitative determination of metabolites by FTIR. Determination of metabolites requires the application of other methods capable of separating the components from the complex matrix and providing structural information as well. Therefore, we extended the investigations to determine the key metabolites by using GC-MS.

3.8. Gas Chromatography Coupled with Mass Spectrometry (GC-MS)

As information provided by FTIR about the chemical changes of the biopolymer bags during composting was limited, a GC-MS analytical method has also been developed to characterize the decomposition process. The goal was to detect the end of the degradation or to find marker molecules to follow the process by a more specific and selective analytical method. Determination of target components by GC-MS requires derivatization to improve their volatility and detectability. Recently, GC-MS methods applying high temperature (pyrolysis, thermal extraction desorption GC-MS) were published to identify markers of different microplastics in the environmental matrices [81] and to study the thermal decomposition of biodegradable polymers, including PBAT [81–83]. Derivatizing agents were rarely utilized in investigations of pyrolytic products. Among others, hexamethyl disilazane (HMDA) proved to be efficient for the chemical modification of alcoholic and carboxylic active hydrogens [83], and earlier, it has been applied to enhance the detection capability of hemp fibers [82]. HMDA is a weak silylating agent. But under pyrolysis conditions, the reaction is fast, as the usual procedure requires 70 °C and longer reaction time. In our study, we chose BSTFA for these purposes because this reagent is more reactive than HMDA. In addition to BSTFA, trimethyl orthoacetate (TMOA) was tested as a derivatizing agent. A neat TMOA reagent was applied at 60 °C for dried residues, but the signal intensities and probably the conversions were not satisfactory. In contrast, silylation resulted in the desired derivatives within 30 min without heating. For the extraction, acetonitrile was as effective as the dichloromethane; thus, it was chosen as a solvent. Typical chromatograms (Figure 11) of the prepared samples contain silyl derivatives of the corresponding monomers (AA, BDO, and PTA). These final metabolites could easily be identified on their mass spectra stored in the NIST database. In addition, three further molecules have been detected, which originate from PBAT polymer and are intermediate metabolites. Two of them have a very similar structure to AA, but mass spectra (Figure 12) contained fragments of higher masses (e.g., 347 or 359) as well, indicating the presence of one (AA+) or two butanediol units (AA++) in the molecule. The third one had similar fragments to the PTA derivative, but the fragmentation pattern regarding the intensities was somewhat different. Fragments of higher masses (e.g., $m/z = 367$ $[M\text{-}15]^+$) also appeared in the mass spectrum, which belongs to the structure (PTA+) consisting of PTA and a butanediol unit as well. As is characteristic for silyl derivatives, molecular ion was not detected, and only the fragments with a methyl group loss appear at low intensities. Samples taken in the first period of composting (2, 6, and 12 weeks) also contained silylated sugars (retention time range, RT = 12–16 min), which may originate from the hydrolysis of starch and the subsequent breakdown of complex carbohydrates into simpler sugars by microbial activity. Silylated glycerol was also detected (RT = 9.88 min), as well as acetyl tributyl citrate in more samples. The observed pattern of decomposition products is substantially different from those obtained under pyrolysis conditions. At 600 °C, the formation of butenyl esters (alpha cleavage and hydrogen transfer) as well as cyclic oligomers are typical, and some of them where proposed as markers to identify polymers in the environmental microplastic [81]. Among the silylated pyrolysate products of PBAT, benzoic acid produced the most intensive peak [83], as decarboxylation of acids also often occurs. Under composting conditions, the original monomers dominate among the decomposition products, and three additional butandiol esters of corresponding acids (AA+, PTA+, and AA++) were also detected (Figures 11–14).

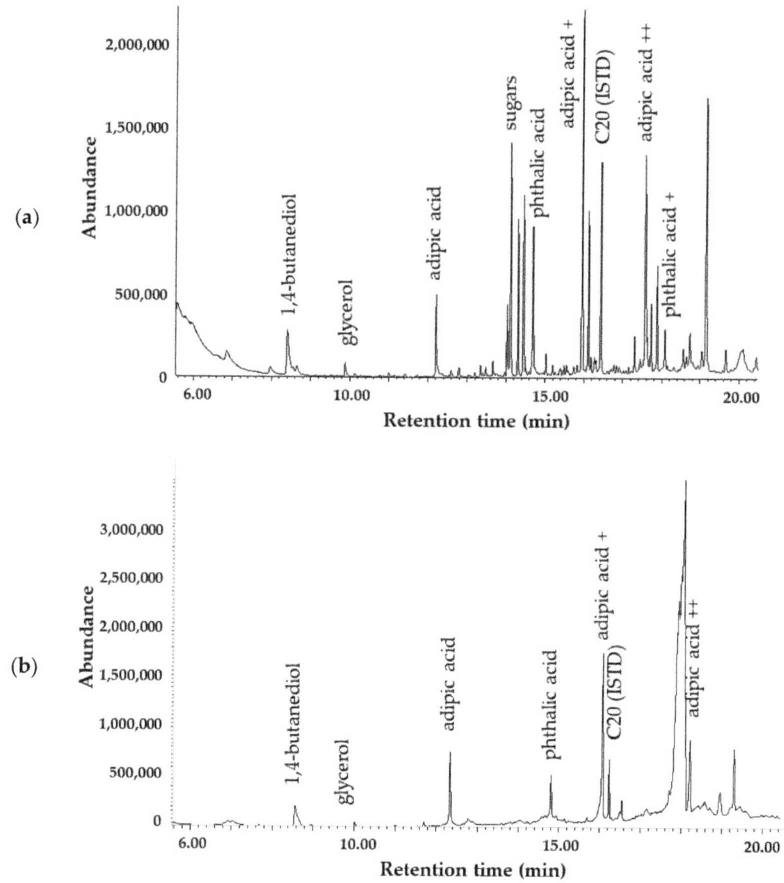

Figure 11. GC-MS chromatogram of the sample showing the distribution and abundance of key compounds after 2 weeks of composting (**a**) and after 1-year maturation (**b**).

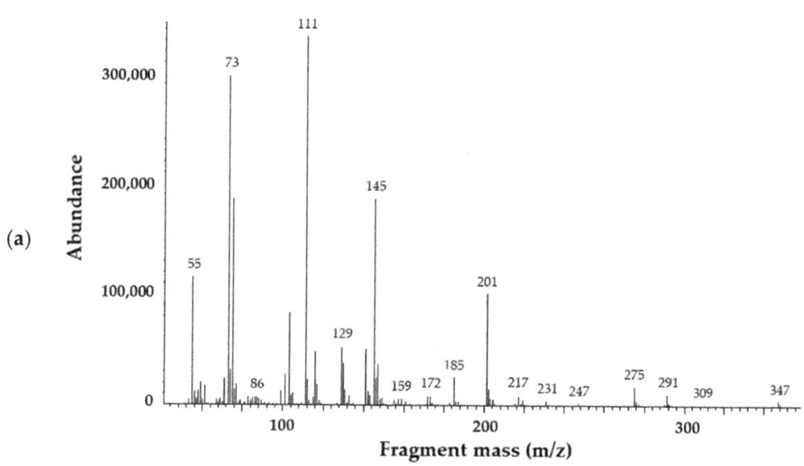

Figure 12. *Cont.*

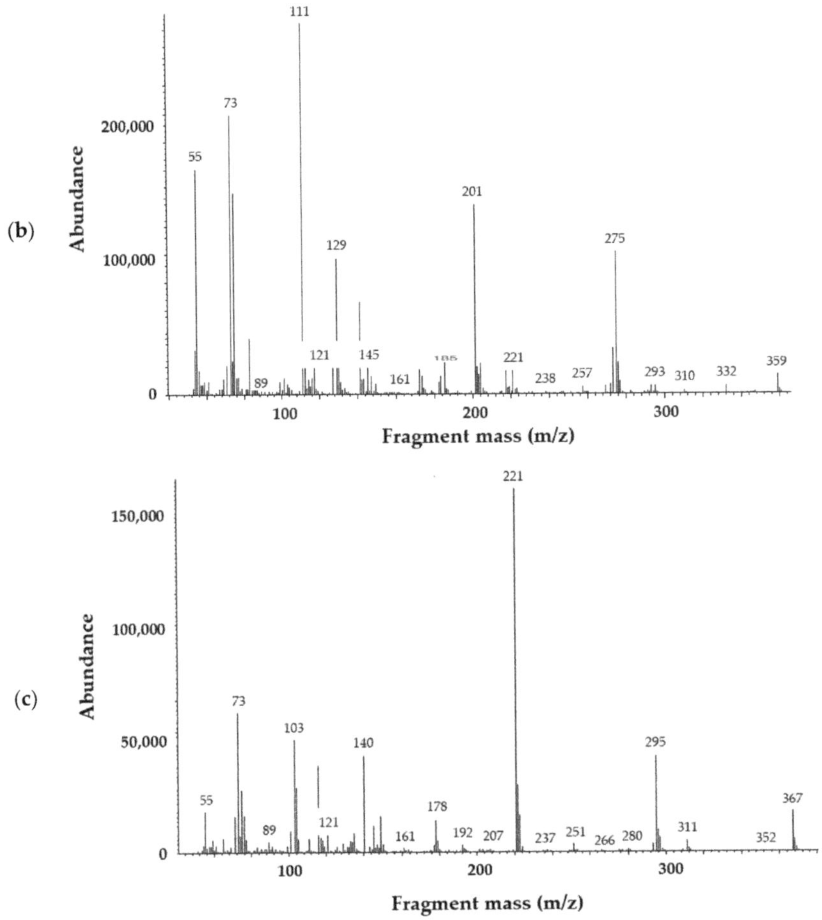

Figure 12. Mass spectra recorded for bis(trimethylsilyl) derivatives of the intermediate metabolites 4-hydroxybutyl)adipate (AA+) (**a**), bis(4-hydroxybutyl)adipate (AA++) (**b**), and (4-hydroxybutyl)terephalate (PTA+) (**c**).

There were no significant differences in the chromatograms (Figure 13) as well as in the levels of decomposition products measured for the smaller (<2 mm) or bigger (>2 mm) particles. Intensities were low after 2 weeks, but there was a significant increase by the 6th week in the concentrations of key metabolites (BDO, AA, and PTA) and their derivatives (AA+ and PTA+), which is related to rapid biodegradation during the thermophilic phase. This observation is in accordance with the FTIR results. Subsequently, a slight drop occurred by the 12th week in the concentration of certain degradation products (e.g., BDO), yet their residual concentrations remained relatively high. This suggests that active composting was still taking place, and microbial degradation continually progressed even at this late stage.

Low levels (<28 μg mL^{-1} in the extract equal to 280 μg g^{-1} in the compost) were determined for PTA, and time-dependent changes showed a slightly increasing trend (Figure 14). Intensity for AA increased from low levels (15 and 7 μg mL^{-1}) after 2 weeks of composting to 131–144 μg mL^{-1} at the 6th week, but then did not alter significantly until the 12th week. Levels of AA (<152 μg mL^{-1}) were generally lower compared to the BDO (<535 μg mL^{-1}). The concentration of BDO reached a maximum in the sixth week for a bigger fraction (>2 mm) and seemed to stabilize after the sixth week for a smaller fraction

(<2 mm). However, the standard deviations calculated from the measured data for the three replicates are generally high; therefore, the results often do not differ significantly from each other. The observed signal intensities of the other three identified molecules (AA+, AA++, and PTA+), containing the corresponding acid and additional butoxy unit(s), were somewhat lower but still in the same orders of magnitude as for the monomers (AA, BDO, and PTA). AA+ resulted in the highest intensity, followed by PTA+, and amounts of AA++ were sometimes low and overlapped with a broad signal. However, PTA+ was more intensive compared to PTA, and its levels were similar in both size fractions. Its estimated concentration in the extract did not increase after the sixth week but remained in the range between 121 and 145 μg mL^{-1}. The same has been observed for A+. For the larger fraction, its level seemed to be lower in the 12th week compared to the 6th week samples, but the difference is not significant because of the high standard deviation values. Due to the interferences, the results for AA++ seem to be not reliable in some cases. Despite the fact that $m/z = 111$ was selected for the estimation of AA++ quantity, which was not a characteristic fragment ion in the mass spectrum of the interfering component(s), its effect could not be eliminated in some cases.

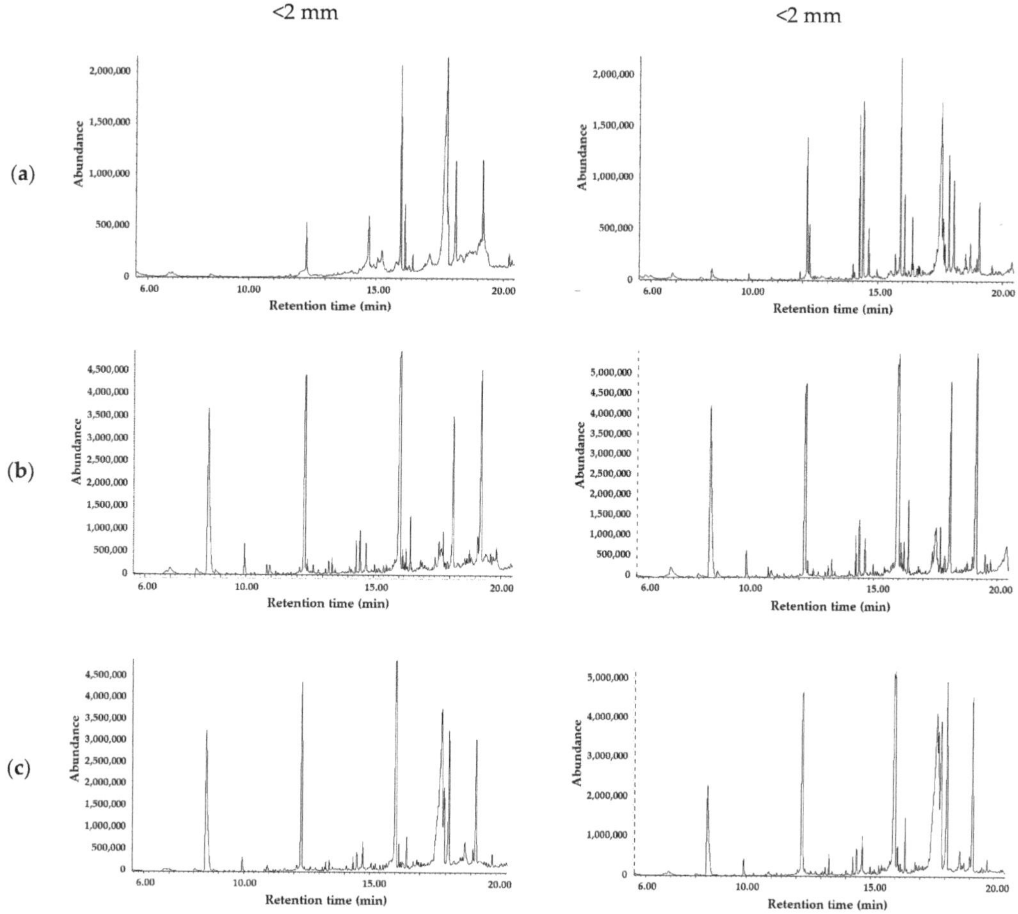

Figure 13. GC-MS chromatograms for <2 mm and >2 mm fractions at 2 (**a**), 6 (**b**), and 12 (**c**) weeks.

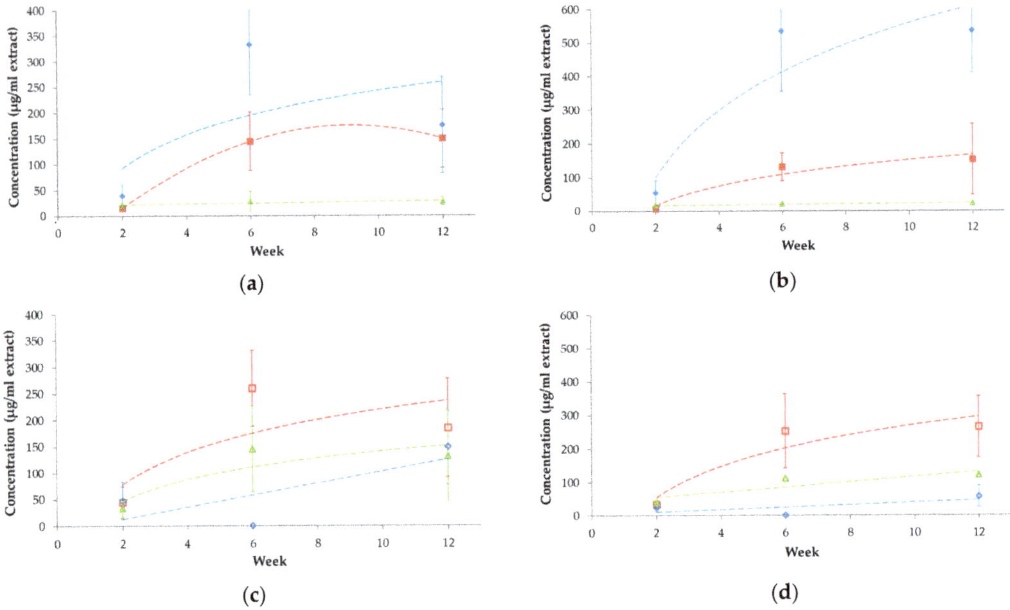

Figure 14. Concentrations in the polybutylene adipate terephthalate-based bioplastic bag fragments from >2 mm (**a**,**b**) and <2 mm (**c**,**d**) compost fraction observed at the 2nd, 6th, and 12th weeks of composting experiment. (**a**,**b**) shows the levels of AA (■ red) and PTA (▲ green). 1,4-butanediol (♦ blue). (**c**,**d**) present the estimated levels for AA+ (□ red), PTA+ (Δ green), and AA++ (◊ blue) based on $m/z = 111$.

Despite the thorough mixing of the composted material, high standard deviation values observed, typically around 50%, indicate that the compost was not so homogeneous. This shows that it is difficult to reach homogeneity in composting performed on the industrial scale. Local conditions, including temperature inhomogeneities in the different parts of compost, result in differences in the decomposition rates; crystallization may also occur, resulting in changes in chemical composition.

Matured 1-year-old compost sample (Figure 11b) did not contain sugars, indicating the successful degradation of starch present in the biopolymer and that of other complex carbohydrates from the manure. Although the sugars were entirely broken down, all other compounds related to the decomposition of PBAT and mentioned above were still detected. However, their intensities were similar to the values determined for second-week samples, typically about one order of magnitude lower (<35 µg mL^{-1}) compared to the samples taken after the 6th or 12th week of composting. Concentrations were 19.3, 3.1, and 14.6 µg mL^{-1} for BDO, AA, and PTA, respectively. Estimated values for AA+, PTA+, and AA++ were 27.5, 29.1, and 31.2 µg mL^{-1}, which means about 275 to 312 µg g^{-1} in the compost. The presence of the decomposition products indicates that the process did not go into completeness.

Although toxicology data are for the monomers, the main decomposition products are favorable as they are regarded as not hazardous to the aquatic environment (e.g., upon leaching from the compost), but the presence of large amounts of microplastic in the environment is of concern. It is worth noting that microplastics are able to bind pollutants (e.g., PAHs, organochlorine pesticides), which can be transported to surface and groundwater [84]. Based on the material safety data sheets, AA has an acute toxicity value for aquatic invertebrates of 46 mg L^{-1}, whereas its chronic LC_{50} value is 17.6 mg L^{-1}. The acute and chronic EC_{50} values for aquatic invertebrates determined for PTA are

640 mg L^{-1} and 42 mg L^{-1}, respectively. Among these compounds, BDO was found to be the least toxic, with an acute LC_{50} value of 813 mg L^{-1} for *D. magna*.

3.9. Ecotoxicity Tests

The ecotoxicity tests were carried out to determine the influence of compost, when mixed at a 50:50 ratio with potting soil, on the germination and growth of three distinct plant species. The results are presented in Table 6. As seen from the instrumental analytical (FTIR, GC-MS) results (see Sections 3.7 and 3.8), PBAT has not fully degraded during the composting process. The ecotoxicity tests can indicate what effect the remaining PBAT may have on plant germination.

Table 6. Effects of the composted biopolymer on germination dynamics determined in standard test protocols using common indicator plant organisms.

Test Species	Treatment [a]	Duration	No. of Seeds	Seeds Germinated	Biomass	Germination Rate		Test Standard
						by Seed Number	by Biomass	
		(Day)		(pc)	(g)	(%)		
Chinese cabbage (*Brassica rapa* ssp. *pekinensis*)	C	5 and 37	20	18.25 ± 2.36	65.37 ± 6.47			EN 16086-1 [52]
	T			16.50 ± 1.29	60.94 ± 17.78	90.4	93.4	
Spring barley (*Hordeum vulgare*)	C	5 and 16	20	18.75 ± 0.96	12.48 ± 0.48			EN 16086-1 [52]
	T			18.50 ± 1.91	8.38 ± 0.59	98.7	67.2	
White mustard (*Sinapis alba*)	C	16	25	25	10.91 ± 0.87			MSZ 08-0012-4:1979 [58] [b]
	T			25	5.44 ± 0.37	100	49.9	

[a] C: control experiments carried out in potting soil; T: treatment experiments carried out in compost mixed at a 50:50 ratio with potting soil; [b] discontinued national test standard.

Chinese cabbage (*Brassica rapa* ssp. *pekinensis*) exhibited a more resilient response to the compost mixture. With a mere 7% germination inhibition, it managed to produce a green mass corresponding to 93% of the control's yield. This suggests that the compost constituents have only a negligible inhibitory effect on the germination and subsequent growth of Chinese cabbage. Consequently, a 50:50 compost-to-soil ratio might be an appropriate medium for fostering the growth of this plant species. However, the assessment for spring barley (*Hordeum vulgare*) highlighted a tangible effect on germination, with the compost causing a germination inhibition of about 33%. This can be deduced from the recorded 67% green mass in the presence of the compost relative to the control sample, which only contained potting soil.

In contrast, the white mustard (*Sinapis alba*) displayed a pronounced sensitivity to the compost mixture, with its germination inhibition of 61%. Such a contrast to the Chinese cabbage implies species-specific responses to the compost mix, possibly due to distinct physiological attributes or growth requirements. Further characterization of the mature compost was performed to understand its phytotoxicity. Its electrical conductivity was recorded at 1251 s cm^{-1} at a temperature of 22.8 °C, and it manifested a relatively low salt content of 0.6%. In plant growth, the concentration of salt can be a pivotal factor, as elevated levels might impede germination. However, in this case, the salt content was deemed non-inhibitory for the studied plants. Another pertinent observation was the respiration intensity of the matured compost, measured at 4 mg O_2 g^{-1} d.m. This low value is indicative of its stability, suggesting that the microbial activity within the compost has seized. Thus, the results suggested the maturity and the non-phytotoxicity of the compost.

The Chinese cabbage tests, showing only 7% germination inhibition, parallels the study by Vaverková and Adamcová (2015) [85], which observed high germination capacities in plant growth tests for compost. These collective results hint at the potential suitability of certain composts for specific crops, like Chinese cabbage. However, the pronounced sensitivity of white mustard to our compost mix, registering a 61% germination inhibition, underscores the intricacies of plant–compost interactions.

Germination rates were calculated based on the number of seeds germinated and the biomasses of the germinated plantlets (as compared to the corresponding controls). It is seen that germination rates based on the number of seeds germinated have always been above 90%; meanwhile, those calculated based on biomasses could be found substantially lower, down to as low as 50%. This practically means that while the compost residue of the PBAT-based bioplastic would not significantly inhibit germination on the basis of seeds germinated, even if the compost was added to the potting soil at a ratio of 1:1, it would still affect germination by resulting in scrubby plantlets compared to the corresponding control. This alerts us that more sophisticated standard test procedures should be established for ecotoxicity assessment for rating germination inhibition.

4. Conclusions

The 90-day composting experiment aimed at investigating the degradation dynamics of PBAT-based bioplastic bags in industrial conditions concluded with several key findings. The findings not only underscore the potential of composting as a viable solution for biodegradable plastic waste management but also emphasize the importance of improving composting parameters to achieve optimal results. The evidence suggests that, under appropriate conditions, PBAT-based bioplastics can be effectively composted, thereby mitigating their environmental impact. However, the study also underscores the importance of ongoing monitoring and optimization.

The initial C/N ratio of 35.6 ± 5.3 was slightly above the maximum value of the literature recommendations. While the EN14045:2003 [29] standard suggests an optimal range for efficient composting, the experiment began with a C/N ratio outside this ideal range. Despite this, there was a substantial decline in the ratio, ending at 15.9 ± 0.3, which is closer to the values reported by Itävaara et al. (1997) [60]. Similarly, the MC showed a significant decrease from 46.08% ± 1.29 to 35.90% ± 1.3 during the high-temperature phase of the composting process. This rapid decline in moisture, coupled with sustained temperatures above 50 °C, likely led to the dehydration of the compost pile. Such dehydration could have hindered the proper degradation of materials, an aspect that should be carefully managed.

Although the degree of disintegration reached an impressive 95% ± 1.86 by the 12th week, the visual examination of the compost revealed a complex matrix. This complexity suggests that while disintegration was successful, the compost still contained intricate material structures that might need further degradation. From the FTIR and extract composition results, it can be concluded that compared with starch, PBAT does not fully degrade. The results of GC-MS analysis of the composted biopolymer indicated that PBAT has partially degraded to intermediate metabolites (4-hydroxybutyl)adipate (AA+), bis(4-hydroxybutyl)adipate (AA++), and (4-hydroxybutyl)terephalate (PTA+). Ecotoxicity tests showed interactions between compost and plant species. While some plants, such as Chinese cabbage, demonstrated positive results to the compost, others, like white mustard, displayed pronounced sensitivity. Such disparities underscore the importance of tailoring compost formulations based on the intended agricultural application. Morphological evaluations, both through optical microscopy and SEM, provided comprehensive insights into the intricate microstructural alterations bioplastics undergo during composting.

Chemical composition changes, as revealed through FTIR spectroscopy and GC-MS, validated the degradation pathways. The FTIR spectra provided insights into the biopolymer's hydrolytic and enzymatic degradation, especially the breakdown of ester linkages within PBAT and the hydrolysis of starch. GC-MS analysis further corroborated these findings, with significant molecular shifts observed throughout the composting process. The experiment also revealed the presence and persistence of terephthalic acid in compost.

Based on the GC-MS results, we can conclude that the PBAT biopolymer is not fully decomposed. Even particles with a size above 2 mm are still present in the compost after the 12th week. As expected, the decomposition products appeared in the samples taken

in the first period. In addition to monomers, more complex molecules (acid butanediol esters) have also been detected. These products did not disappear from the composted material after 1 year; thus, neither of them was a suitable marker to indicate the end of degradation. The observed differences among the parallel samples confirm that the rate of decomposition is strongly influenced by the local conditions in the compost, including the temperature. Under favorable conditions, the crystallization of amorphous PBAT polymer may lead to a more stable phase, which is prone to resist biodegradation. Disintegration of the biopolymers results in microplastic, which may remain in the compost for a longer period of time.

For future studies, it is recommended to focus not only on the general compost matrix but also on the finer fractions, particularly those less than 2 mm. Additionally, careful examination of microplastics in the compost should be conducted. This will ensure a comprehensive understanding of the degradation process and the environmental safety of the resulting compost. Moreover, it is crucial to optimize composting conditions, especially MC and temperature, to prevent dehydration and ensure effective degradation. The role of compost additives and the persistence of specific compounds like terephthalic acid warrant further investigation. Additionally, a more in-depth analysis of compost fractions and the potential generation of microplastics will be essential for assessing the environmental impact of composting bioplastics like the PBAT-based bioplastic studied. The results of the current study underline above all that decomposition of the PBAT polymer has not been completed during the study period, while according to the current guidelines, it could be considered decomposed based on both visual tools for mechanical fragmentation and germination tests. This indeed raises the need for more precautious governance and further assessment of the bioplastic decomposition process.

It would also be beneficial to explore the effects of different pre-treatment methods on the composting process. It would be worthwhile to establish a requirement for analyzing the compost for potentially toxic compounds and ensuring that it is safe for soil application in agricultural fields. Furthermore, understanding the long-term effects of compost derived from bioplastics on soil health, fertility, and crop yield is essential. For this reason, in composting plants where biopolymers are also treated, the examination of microplastics must be recommended during the authorization procedure. However, challenges in measuring microplastics in a solid matrix and the shortcomings of current standards need to be assessed. This work can significantly increase our understanding of the intricate microplastic–compost interactions and highlights the potential analytical methods of quantification of said components. To conclude, studying the composability of bioplastics under industrial conditions is vital because, as the world grapples with plastic pollution, bioplastics present a sustainable alternative. Ensuring their safe and efficient composting in industrial settings paves the way for a circular economy, where materials are continually reused and recycled, minimizing the environmental impact.

Author Contributions: Conceptualization, Z.I.V., A.S. and L.A.; methodology, M.M. and Z.I.V.; validation, M.M., M.D. and Z.I.V.; formal analysis, M.M., M.D., M.G., Z.I.V., G.F., T.K. and Á.R.; investigation, M.M., M.D. and Z.I.V.; writing—original draft preparation, M.M., M.D. and A.S.; writing—review and editing, M.G., Z.I.V. and L.A.; supervision, A.S. and L.A.; project administration, Z.I.V.; funding acquisition, A.S. and L.A. All authors have read and agreed to the published version of the manuscript.

Funding: This work was funded by the Hungarian National Research, Development and Innovation Office within project 2022-2.1.1-NL-2022-00006, "Development of the Agrotechnology National Laboratory" (grant agreement NKFIH-3524-1/2022), supported by the National Research, Development and Innovation Fund by the Hungarian Ministry of Culture and Innovation, within the project, "The feasibility of the circular economy during national defense activities", of the 2021 Thematic Excellence Program (grant no. TKP2021-NVA-22), led by the Centre for Circular Economy Analysis. Research funding was also provided by the Flagship Research Groups Program 2024 of the Hungarian University of Agriculture and Life Sciences.

Institutional Review Board Statement: Not applicable.

Data Availability Statement: Data are contained within the article.

Acknowledgments: The authors thank the Centre for Circular Economy Analysis, founded by the Hungarian University of Agricultural and Life Sciences (MATE), for enrolling this project in its education profile.

Conflicts of Interest: Authors Mariem Damak, Zsolt István Varga, György Fekete, and László Aleksza were employed by the company Profikomp Environmental Technologies Inc. The remaining authors declare that the research was conducted in the absence of any commercial or financial relationships that could be construed as a potential conflict of interest.

References

1. Üveges, Z.; Damak, M.; Klátyik, S.; Ramay, M.W.; Fekete, G.; Varga, Z.; Gyuricza, C.; Székács, A.; Aleksza, L. Biomethane potential in anaerobic biodegradation of commercial bioplastic materials. *Fermentation* **2023**, *9*, 261. [CrossRef]
2. Statista. *Annual Production of Plastics Worldwide from 1950 to 2023*; Statista Research Department: New York, NY, USA, 2024; Available online: https://www.statista.com/statistics/282732/global-production-of-plastics-since-1950 (accessed on 25 November 2024).
3. OECD. *Policy Scenarios for Eliminating Plastic Pollution by 2040*; OECD: Paris, France, 2024; Available online: https://www.oecd.org/en/publications/policy-scenarios-for-eliminating-plastic-pollution-by-2040_76400890-en.html (accessed on 25 November 2024). [CrossRef]
4. OECD. *Increased Plastic Leakage and Greenhouse Gas Emissions*; OECD: Paris, France, 2022; Available online: https://www.oecd.org/environment/plastics/increased-plastic-leakage-and-greenhouse-gas-emissions.htm (accessed on 25 November 2024).
5. Cabernard, L.; Pfister, S.; Oberschelp, C.; Hellweg, S. Growing environmental footprint of plastics driven by coal combustion. *Nat. Sustain.* **2021**, *5*, 139–148. [CrossRef]
6. Liu, L.; Xu, M.; Ye, Y.; Zhang, B. On the degradation of (micro)plastics: Degradation methods, influencing factors, environmental impacts. *Sci. Total Environ.* **2022**, *806*, 151312. [CrossRef] [PubMed]
7. Bostan, N.; Ilyas, N.; Akhtar, N.; Mehmood, S.; Saman, R.U.; Sayyed, R.Z.; Shatid, A.A.; Alfaifi, M.Y.; Elbehairi, S.E.I.; Pandiaraj, S. Toxicity assessment of microplastic (MPs); a threat to the ecosystem. *Environ. Res.* **2023**, *234*, 116523. [CrossRef]
8. Zeb, A.; Liu, W.; Ali, N.; Shi, R.; Wang, Q.; Wang, J.; Li, J.; Yin, C.; Liu, J.; Yu, M.; et al. Microplastic pollution in terrestrial ecosystems: Global implications and sustainable solutions. *J. Hazard. Mater.* **2024**, *461*, 132636. [CrossRef]
9. Ali, N.; Khan, M.H.; Ali, M.; Sidra; Ahmad, S.; Khan, A.; Nabi, G.; Ali, F.; Bououdina, M.; Kyzas, G.Z. Insight into microplastics in the aquatic ecosystem: Properties, sources, threats and mitigation strategies. *Sci. Total Environ.* **2024**, *913*, 169489. [CrossRef]
10. Martínez, A.; Perez-Sanchez, E.; Caballero, A.; Ramírez, R.; Quevedo, E.; Salvador-García, D. PBAT is biodegradable but what about the toxicity of its biodegradation products? *J. Mol. Model.* **2024**, *30*, 273. [CrossRef]
11. Dissanayake, P.D.; Withana, P.A.; Sang, M.K.; Cho, Y.; Park, J.; Oh, D.X.; Chang, S.X.; Lin, C.S.K.; Bank, M.S.; Hwang, S.Y.; et al. Effects of biodegradable poly(butylene adipate-co-terephthalate) and poly(lactic acid) plastic degradation on soil ecosystems. *Soil Use Manag.* **2024**, *40*, e13055. [CrossRef]
12. European Bioplastics. What Are Bioplastics. Available online: https://www.european-bioplastics.org/bioplastics (accessed on 25 November 2024).
13. Adekiya, A.O.; Ejue, W.S.; Olayanju, A.; Dunsin, O.; Aboyeji, C.M.; Aremu, C.; Adegbite, K.; Akinpelu, O. Different organic manure sources and NPK fertilizer on soil chemical properties, growth, yield and quality of okra. *Sci. Rep.* **2020**, *10*, 16083. [CrossRef]
14. Abraham, A.; Park, H.; Choi, O.; Sang, B. Anaerobic co-digestion of bioplastics as a sustainable mode of waste management with improved energy production—A review. *Biores. Technol.* **2021**, *322*, 124537. [CrossRef]
15. Mendes, A.C.; Pedersen, G.A. Perspectives on sustainable food packaging—Is bio-based plastics a solution? *Trends Food Sci. Technol.* **2021**, *112*, 839–846. [CrossRef]
16. Havstad, M.R. Biodegradable plastics. In *Plastic Waste and Recycling*; Letcher, T.M., Ed.; Elsevier: London, UK, 2020; pp. 97–129. [CrossRef]
17. Emadian, S.M.; Onay, T.T.; Demirel, B. Biodegradation of bioplastics in natural environments. *Waste Manag.* **2017**, *59*, 526–536. [CrossRef] [PubMed]
18. Nair, N.R.; Sekhar, V.C.; Nampoothiri, K.; Pandey, A. Biodegradation of biopolymers. In *Current Developments in Biotechnology and Bioengineering. Production, Isolation and Purification of Industrial Products*; Pandey, A., Negi, S., Soccol, C.R., Eds.; Elsevier: Amsterdam, The Netherlands, 2017; pp. 739–755. [CrossRef]
19. Polman, E.M.N.; Gruter, G.-J.M.; Parsons, J.R.; Tietema, A. Comparison of the aerobic biodegradation of biopolymers and the corresponding bioplastics: A review. *Sci. Total Environ.* **2017**, *753*, 141953. [CrossRef] [PubMed]
20. Müller, R. Biodegradability of polymers: Regulations and methods for testing. In *Biopolymers Online: Biology, Chemistry, Biotechnology, Applications*; Wiley-VCH Verlag: Weinheim, Germany, 2005. [CrossRef]

21. Ismail, M.; Abouhmad, A.; Warlin, N.; Pyo, S.-H.; Örn, O.E.; Al-Rudainy, B.; Tullberg, C.; Zhang, B.; Hatti-Kaul, R. Closing the loop for poly(butylene-adipate-co-terephthalate) recycling: Depolymerization, monomers separation, and upcycling. *Green Chem.* **2024**, *26*, 3863–3873. [CrossRef]
22. Wang, L.; Chang, R.; Ren, Z.; Meng, X.; Li, Y.; Gao, M. Mature compost promotes biodegradable plastic degradation and reduces greenhouse gas emission during food waste composting. *Sci. Total Environ.* **2024**, *926*, 172081. [CrossRef]
23. Azim, K.; Soudi, B.; Boukhari, S.; Perissol, C.; Roussos, S.; Alami, I.T. Composting parameters and compost quality: A literature review. *Org. Agric.* **2017**, *8*, 141–158. [CrossRef]
24. Sánchez, Ó.J.; Ospina, D.A.; Montoya, S. Compost supplementation with nutrients and microorganisms in composting process. *Waste Manag.* **2017**, *69*, 136–153. [CrossRef]
25. Mehta, C.; Palni, U.; Franke-Whittle, I.; Sharma, A. Compost: Its role, mechanism and impact on reducing soil-borne plant diseases. *Waste Manag.* **2014**, *34*, 607–622. [CrossRef]
26. *EN 13432:2001*; Packaging—Requirements for Packaging Recoverable through Composting and Biodegradation—Test Scheme and Evaluation Criteria for the Final Acceptance of Packaging. European Committee for Standardization: Brussels, Belgium, 2001. Available online: https://docs.european-bioplastics.org/publications/bp/EUBP_BP_En_13432.pdf (accessed on 25 November 2024).
27. *EN 17033:2018*; Plastics—Biodegradable Mulch Films for Use in Agriculture and Horticulture—Requirements and Test Methods. European Committee for Standardization: Brussels, Belgium, 2018. Available online: https://www.en-standard.eu/une-en-17033-2018-plastics-biodegradable-mulch-films-for-use-in-agriculture-and-horticulture-requirements-and-test-methods (accessed on 25 November 2024).
28. *EN 14995:2006*; Plastics—Evaluation of Compostability—Test Scheme and Specifications. European Committee for Standardization: Brussels, Belgium, 2006. Available online: https://www.en-standard.eu/bs-en-14995-2006-plastics-evaluation-of-compostability-test-scheme-and-specifications (accessed on 25 November 2024).
29. *EN 14045:2003*; Packaging—Evaluation of the Disintegration of Packaging Materials in Practical Use under Composting Conditions—Test Scheme and Requirements. European Committee for Standardization: Brussels, Belgium, 2003. Available online: https://cdn.standards.iteh.ai/samples/13123/7038bd7bdf7a4281a99eb8e9b2a31fce/SIST-EN-14045-2003.pdf (accessed on 25 November 2024).
30. *EN 14806:2005*; Packaging—Preliminary Evaluation of the Disintegration of Packaging Materials under Simulated Composting Conditions in a Laboratory Scale Test. European Committee for Standardization: Brussels, Belgium, 2005. Available online: https://www.en-standard.eu/bs-en-14806-2005-packaging-preliminary-evaluation-of-the-disintegration-of-packaging-materials-under-simulated-composting-conditions-in-a-laboratory-scale-test (accessed on 25 November 2024).
31. *ISO 17088:2021*; Specifications for Compostable Plastics. International Organization for Standardization: Geneva, Switzerland, 2021. Available online: https://www.iso.org/standard/74994.html (accessed on 25 November 2024).
32. *ISO 16929:2021*; Plastics—Determination of the Degree of Disintegration of Plastic Materials Under Defined Composting Conditions in a Pilot-Scale Test. International Organization for Standardization: Geneva, Switzerland, 2021. Available online: https://www.iso.org/standard/80302.html (accessed on 25 November 2024).
33. *ISO 18606:2013*; Packaging and the Environment—Organic Recycling. International Organization for Standardization: Geneva, Switzerland, 2013. Available online: https://www.iso.org/standard/55874.html (accessed on 25 November 2024).
34. *ISO 20200:2023*; Plastics—Determination of the Degree of Disintegration of Plastic Materials under Simulated Composting Conditions in a Laboratory-Scale Test. International Organization for Standardization: Geneva, Switzerland, 2023. Available online: https://www.iso.org/standard/81932.html (accessed on 25 November 2024).
35. *ISO 17556:2019*; Plastics—Determination of the Ultimate Aerobic Biodegradability in Soil by Measuring the Oxygen Demand in a Respirometer or the Amount of Carbon Dioxide Evolved. International Organization for Standardization: Geneva, Switzerland, 2019. Available online: https://www.iso.org/standard/74993.html (accessed on 25 November 2024).
36. *ASTM D6400:2023*; Standard Specification for Labeling of Plastics Designed to be Aerobically Composted in Municipal or Industrial Facilities. ASTM International: West Conshohocken, PA, USA, 2023. Available online: https://www.astm.org/d6400-23.html (accessed on 25 November 2024).
37. *ASTM D6868:2021*; Standard Specification for Labeling of End Items that Incorporate Plastics and Polymers as Coatings or Additives with Paper and Other Substrates Designed to be Aerobically Composted in Municipal or Industrial Facilities. ASTM International: West Conshohocken, PA, USA, 2021. Available online: https://www.astm.org/d6868-21.html (accessed on 25 November 2024).
38. *AS 43736:2006*; Plastics—Biodegradable Plastics Suitable for Composting and Other Microbial Treatment. Standards Australia: Sydney, Australia, 2006. Available online: https://www.saiglobal.com/pdftemp/previews/osh/as/as4000/4700/4736-2006.pdf (accessed on 25 November 2024).
39. *EN 14046:2003*; Evaluation of the Ultimate Aerobic Biodegradability and Disintegration of Packaging Materials under Controlled Composting Conditions—Method by Analysis of Released Carbon Dioxide. European Committee for Standardization: Brussels, Belgium, 2003. Available online: https://i2.saiglobal.com/mpc2v/preview/98691170373.pdf?sku=860959_SAIG_NSAI_NSAI_2048280&nsai_sku=i-s-en-14046-2003-860959_saig_nsai_nsai_2048280 (accessed on 25 November 2024).
40. Borchani, K.E.; Carrot, C.; Jaziri, M. Biocomposites of Alfa fibers dispersed in the Mater-Bi® type bio-plastic: Morphology, mechanical and thermal properties. *Compos. Part A Appl. Sci. Manuf.* **2015**, *78*, 371–379. [CrossRef]

41. *ASTM D1003:2021*; Standard Test Method for Haze and Luminous Transmittance of Transparent Plastics. ASTM International: West Conshohocken, PA, USA, 2021. Available online: https://www.astm.org/d1003-21.html (accessed on 25 November 2024).
42. Material Data Center. Material Data Center Datasheet of Mater-Bi NF803—PSAC—Novamont S.p.A [Online]. MDC. 2023. Available online: https://www.materialdatacenter.com/ms/en/Mater-Bi/Novamont+SpA/1784 (accessed on 25 November 2024).
43. *ASTM D3418:2021*; Standard Test Method for Transition Temperatures of Polymers by Differential Scanning Calorimetry. ASTM International: West Conshohocken, PA, USA, 2021. Available online: https://www.astm.org/d3418-21.html (accessed on 25 November 2024).
44. *ASTM D792:2020*; Standard Test Methods for Density and Specific Gravity (Relative Density) of Plastics by Displacement. ASTM International: West Conshohocken, PA, USA, 2020. Available online: https://www.astm.org/standards/d792 (accessed on 25 November 2024).
45. *ASTM D882*; Standard Test Method for Tensile Properties of Thin Plastic Sheeting. ASTM International: West Conshohocken, PA, USA, 2020. Available online: https://www.astm.org/d0882-18.html (accessed on 25 November 2024).
46. Choi, K. Optimal operating parameters in the composting of swine manure with wastepaper. *J. Environ. Sci. Health B* **1999**, *34*, 975–987. [CrossRef]
47. Bueno, P.; Tapias, R.; López, L.; Díaz, M.J. Optimizing composting parameters for nitrogen conservation in composting. *Bioresour. Technol.* **2008**, *99*, 5069–5077. [CrossRef] [PubMed]
48. Haug, R.T. *The Practical Handbook of Compost Engineering*; Lewis Publishers: Boca Raton, FL, USA, 2018. [CrossRef]
49. *MSZ EN ISO 16948:2015*; Solid Biofuels—Determination of Total Content of Carbon, Hydrogen, and Nitrogen—Instrumental Methods. International Organization for Standardization: Geneva, Switzerland, 2015. Available online: https://www.iso.org/standard/58004.html (accessed on 25 November 2024).
50. *MSZ EN ISO 16994:2017*; Solid Biofuels—Determination of Total Content of Sulfur and Chlorine. International Organization for Standardization: Geneva, Switzerland, 2016. Available online: https://www.iso.org/standard/70097.html (accessed on 25 November 2024).
51. *MSZ EN 15935:2012*; Sludge, Treated Biowaste, and Soil—Determination of Loss on Ignition. International Organization for Standardization: Geneva, Switzerland, 2013. Available online: https://standards.iteh.ai/catalog/standards/sist/e2923374-f17f-4e46-8fe5-f0f318857d87/sist-en-15935-2012 (accessed on 25 November 2024).
52. *EN 16086-1:2012*; Soil Improvers and Growing Media—Determination of Plant Response—Part 1: Petri Dish Test Using Cress. European Committee for Standardization: Brussels, Belgium, 2003. Available online: https://www.ivami.com/en/ecotoxicology-algae-i-daphnia-magna-i-and-fish/7253-response-of-plants-in-soil-improvers-and-culture-substrates-part-1-chinese-cabbage-pot-growth-test-en-16086-1-2011-standard (accessed on 25 November 2024).
53. *MSZ EN 13037:2012*; Soil Improvers and Growing Media—Determination of pH. European Committee for Standardization: Brussels, Belgium, 2012. Available online: https://www.en-standard.eu/din-en-13037-soil-improvers-and-growing-media-determination-of-ph/?srsltid=AfmBOop2ZLJ61Fcq0Avqa18OOBm13CLLkvqv34yklOOioPB64c103kbX (accessed on 25 November 2024).
54. *ISO 7888:1985*; Water Quality—Determination of Electrical Conductivity. International Organization for Standardization: Geneva, Switzerland, 1985. Available online: https://www.iso.org/standard/14838.html (accessed on 25 November 2024).
55. *ISO 8502-6:2020*; Preparation of Steel Substrates Before Application of Paints and Related Products—Tests for the Assessment of Surface Cleanliness—Part 6: Extraction of Soluble Contaminants for Analysis—The Bresle Method. International Organization for Standardization: Geneva, Switzerland, 2020. Available online: https://www.iso.org/standard/73862.html (accessed on 25 November 2024).
56. *ISO 8502-9:2020*; Preparation of Steel Substrates Before Application of Paints and Related Products—Tests for the Assessment of Surface Cleanliness—Part 9: Field Method for the Conductometric Determination of Water-Soluble Salts. International Organization for Standardization: Geneva, Switzerland, 2020. Available online: https://www.iso.org/standard/73863.html (accessed on 25 November 2024).
57. Malinska, K. Application of a modified OxiTop respirometer for laboratory composting studies. *Arch. Environ. Prot.* **2016**, *42*, 56–62. [CrossRef]
58. *MSZ 08-0012-4:1979*; Standard for the Determination of Moisture Content. National Accreditation Body: Budapest, Hungary, 1979.
59. Flavel, T.C.; Murphy, D.V. Carbon and nitrogen mineralization rates after application of organic amendments to soil. *J. Environ. Qual.* **2006**, *35*, 183–193. [CrossRef] [PubMed]
60. Itävaara, M.; Vikman, M.; Venelampi, O. Windrow Composting of biodegradable packaging materials. *Compos. Sci. Util.* **1997**, *5*, 84–92. [CrossRef]
61. Kianirad, M.; Muazardalan, M.; Savaghebi, G.; Farahbakhsh, M.; Mirdamadi, S. Effects of temperature treatment on corn cob composting and reducing of composting time: A comparative study. *Waste Manag. Res.* **2009**, *28*, 882–887. [CrossRef]
62. Jian, J.; Xiangbin, Z.; Xianbo, H. An overview on synthesis, properties and applications of poly(butylene-adipate-co-terephthalate)—PBAT. *Adv. Ind. Eng. Poly. Res.* **2020**, *3*, 19–26. [CrossRef]
63. Tester, C.F. Organic amendment effects on physical and chemical properties of a sandy soil. *Soil Sci. Soc. Am. J.* **1990**, *54*, 827–831. [CrossRef]
64. Cronjé, A.L.; Turner, C.; Williams, A.G.; Barker, A.J.; Guy, S. The respiration rate of composting pig manure. *Compos. Sci. Util.* **2004**, *12*, 119–129. [CrossRef]

65. Steiner, C.; Das, K.C.; Melear, N.D.; Lakly, D. Reducing nitrogen loss during poultry litter composting using biochar. *J. Environ. Qual.* **2010**, *39*, 1236–1242. [CrossRef]
66. Kale, G.; Kijchavengkul, T.; Auras, R.; Rubino, M.; Selke, S.; Singh, S.P. Compostability of bioplastic packaging materials: An overview. *Macromol. Biosci.* **2007**, *7*, 255–277. [CrossRef]
67. Tabasi, R.Y.; Ajji, A. Selective degradation of biodegradable blends in simulated laboratory composting. *Polym. Degrad. Stab.* **2015**, *120*, 435–442. [CrossRef]
68. Ruggero, F.; Belardi, S.; Caretti, E.; Lotti, T.; Lubello, C.; Gori, R. Rigid and film bioplastics degradation under suboptimal composting conditions: A kinetic study. *Waste Manag. Res.* **2021**, *40*, 1311–1321. [CrossRef] [PubMed]
69. Ruggero, F.; Carretti, E.; Gori, R.; Lotti, T.; Lubello, C. Monitoring of degradation of starch-based biopolymer film under different composting conditions, using TGA, FTIR and SEM analysis. *Chemosphere* **2020**, *246*, 125770. [CrossRef] [PubMed]
70. Jung, S.; Cho, S.-H.; Kim, K.-H.; Kwon, E.E. Progress in quantitative analysis of microplastics in the environment: A review. *Chem. Eng. J.* **2021**, *422*, 130154. [CrossRef]
71. Fojt, J.; David, J.; Přikryl, R.; Řezáčová, V.; Kučerík, J. A critical review of the overlooked challenge of determining micro-bioplastics in soil. *Sci. Total Environ.* **2020**, *745*, 140975. [CrossRef]
72. Fojt, J.; Románeková, I.; Procházková, P.; David, J.; Brtnický, M.; Kučerík, J. A simple method for quantification of polyhydroxybutyrate and polylactic acid micro-bioplastics in soils by evolved gas analysis. *Molecules* **2022**, *27*, 1898. [CrossRef]
73. Ruggero, F.; Gori, R.; Lubello, C. Methodologies to assess biodegradation of bioplastics during aerobic composting and anaerobic digestion: A review. *Waste Manag. Res.* **2019**, *37*, 959–975. [CrossRef]
74. Dammak, M.; Fourati, Y.; Tarrés, Q.; Delgado-Aguilar, M.; Mutjé, P.; Boufi, S. Blends of PBAT with plasticized starch for packaging applications: Mechanical properties, rheological behaviour and biodegradability. *Ind. Crops Prod.* **2020**, *144*, 112061. [CrossRef]
75. Weng, Y.-X.; Jin, Y.-J.; Meng, Q.-Y.; Wang, L.; Zhang, M.; Wang, Y.-Z. Biodegradation behavior of poly(butylene adipate-co-terephthalate) (PBAT), poly(lactic acid) (PLA), and their blend under soil conditions. *Polym. Test.* **2013**, *32*, 918–926. [CrossRef]
76. Ruggero, F.; Onderwater, R.; Carretti, E.; Roosa, S.; Benali, S.; Raquez, J.M.; Gori, R.; Lubello, C.; Wattiez, R. Degradation of film and rigid bioplastics during the thermophilic phase and the maturation phase of simulated composting. *J. Polym. Environ.* **2021**, *29*, 3015–3028. [CrossRef]
77. Myalenko, D.; Fedotova, O. Physical, mechanical, and structural properties of the polylactide and polybutylene adipate terephthalate (PBAT)-based biodegradable polymer during compost storage. *Polymers* **2023**, *15*, 1619. [CrossRef] [PubMed]
78. Mano, J.F.; Koniarova, D.; Reis, R.L. Thermal properties of thermoplastic starch/synthetic polymer blends with potential biomedical applicability. *J. Mater. Sci. Mater. Med.* **2003**, *14*, 127–135. [CrossRef] [PubMed]
79. Al-Itry, R.; Lamnawar, K.; Maazouz, A. Improvement of thermal stability, rheological and mechanical properties of PLA, PBAT and their blends by reactive extrusion with functionalized epoxy. *Polym. Degrad. Stab.* **2012**, *97*, 1898–1914. [CrossRef]
80. Park, J.W.; Im, S.S.; Kim, S.H.; Kim, Y.H. Biodegradable polymer blends of poly(L-lactic acid) and gelatinized starch. *Polym. Eng. Sci.* **2000**, *40*, 2539–2550. [CrossRef]
81. Kittner, M.; Eisentraut, P.; Dittmann, D.; Braun, U. Decomposability versus detectability: First validation of TED-GC/MS for microplastic detection in different environmental matrices. *Appl. Res.* **2024**, *3*, e202200089. [CrossRef]
82. De Falco, F.; Nacci, T.; Durndell, L.; Thompson, R.C.; Degano, I.; Modugno, F. A thermoanalytical insight into the composition of biodegradable polymers and commercial products by EGA-MS and Py-GC-MS. *J. Anal. Appl. Pyrolysis* **2023**, *171*, 105937. [CrossRef]
83. Coralli, I.; Rombolà, A.G.; Fabbri, D. Analytical pyrolysis of the bioplastic PBAT poly(butylene adipate-co-terephthalate). *J. Anal. Appl. Pyrolysis* **2024**, *181*, 106577. [CrossRef]
84. Campanale, C.; Dierkes, G.; Massarelli, C.; Bagnuolo, G.; Uricchio, V.F. A relevant screening of organic contaminants present on freshwater and pre-production microplastics. *Toxics* **2020**, *8*, 100. [CrossRef]
85. Vaverková, M.D.; Adamcová, D. Biodegrability of bioplastic materials in a controlled composting environment. *J. Ecol. Eng.* **2015**, *16*, 155–160. [CrossRef]

polymers MDPI

Article

Microbial Biodegradation of Synthetic Polyethylene and Polyurethane Polymers by Pedospheric Microbes: Towards Sustainable Environmental Management

Maryam Najam [1], Sana Javaid [1], Shazia Iram [1,*], Kingkham Pasertsakoun [2], Marianna Oláh [2], András Székács [2] and László Aleksza [2,3,*]

1 Department of Environmental Sciences, Fatima Jinnah Women University, The Mall, Rawalpindi 46000, Pakistan; maryam.najam@fjwu.edu.pk (M.N.); sana.javaid@fjwu.edu.pk (S.J.)
2 Institute of Environmental Sciences, Hungarian University of Agriculture and Life Sciences, Páter Károly u. 1, H-2100 Gödöllő, Hungary; kingkham.pasertsakoun@phd.uni-mate.hu (K.P.); olah.marianna@uni-mate.hu (M.O.); szekacs.andras@uni-mate.hu (A.S.)
3 Profikomp Environmental Technologies Inc., Kühne Ede u. 7, H-2100 Gödöllő, Hungary
* Correspondence: shaizairam@fjwu.edu.pk (S.I.); aleksza.laszlo@uni-mate.hu (L.A.); Tel.: +92-51-9292900-(2083) (S.I.); +36-28-512-490 (L.A.)

Abstract: This study attempted to isolate and identify pedospheric microbes originating in dumpsites and utilized them for the degradation of selected synthetic polymers for the first time in a cost-effective, ecologically favorable and sustainable manner. Specifically, low-density polyethylene (LDPE) and polyurethane (PUR) were converted by the isolated fungi, i.e., *Aspergillus flavus, A terreus, A. clavatus, A. nigers* and bacterial coccus and filamentous microbes and assessed in a biotransformative assay under simulated conditions. Commendable biodegradative potentials were exhibited by the isolated microbes against polymers that were analyzed over a span of 30 days. Among the selected fungal microbes, the highest activity was achieved by *A. niger*, expressing 55% and 40% conversion of LDPE and PUR, respectively. In the case of bacterial strains, 50% and 40% conversion of LDPE and PUR degradation was achieved by coccus. Fourier transform infrared spectroscopy (FT-IR) and thermogravimetric analysis (TGA) were utilized to analyze the degradative patterns in terms of vibrational and thermal characteristics, and stereomicroscopic analysis was performed for the visual assessment of morphological variations. Profound structural transformations were detected in FT-IR spectra and TGA thermograms for the selected microbes. Stereomicroscopic analysis was also indicative of the remarkable transformation of the surface morphology of these polymers after degradation by microbes in comparison to the reference samples not treated with any pedospheric microbes. The results are supportive of the utilization of the selected pedospheric microbes as environmental remediators for the cleanup of persistent polymeric toxins. This current work can be further extended for the successful optimization of further augmented percentages by using other pedospheric microbes for the successful adoption of these biotechnological tools at a practical level.

Keywords: polyurethane; polyethylene; biodegradation; microbial degradation; soil microbes; sustainability

1. Introduction

Environmental degradation in the current era can be attributed to a number of anthropogenic activities that give rise to either permanent or long-term changes in the ecospheric regime. Such practices are not only harmful towards human beings and other life forms,

but, in an uncontrolled manner, they are destructive to the planet's ecological integrity [1,2]. This includes the utilization of synthetic products in the form of daily use commodities, such as plastic bags, which are known for their persistent or potentially persistent nature [3]. The issue of plastic waste due to human activity is widespread throughout the environment. In 2019 alone, 460 million tons of plastic was produced, and 353 million tons were discarded in a single year. This amount is expected to triple by 2060 [4]. Great efforts have been devoted to the fulfillment of human needs in terms of rigorous research that has been conducted on food and energy [5–8], in addition to the environmental detoxification of different pollutants [6,8–10]. Some of these pollutants require urgent attention due to their toxic and persistent nature, in addition to the constant, numerous forms of damage they cause to humans and other life forms. Polymeric pollution due to plastics is a significant issue faced by humanity in the present era [11]. Life without any aspect of utilization of polymeric substances seems to be unattainable due to the complete domination of those products in a multiplicity of forms [12].

Polymers are formed from monomeric units linked in a repetitive manner, either in a linear or branched morphology [13,14]. The effectiveness and large-scale utilization of polymers can be attributed to their ease of fabrication, stability, flexibility, durability and other favorable physical properties [15,16]. Plastics and relevant products have become an important part of human life. However, the type of pollution associated with such products cannot be ignored. They can enter lithospheric and hydrospheric zones in the form of plastic debris from a wide range of products, such as ropes, plastic bags, fishing nets or other household objects or occupational equipment, and remain there for periods of unprecedented durations. Fifty percent of plastic materials are considered disposable, including utensils, packaging and plastic bags [17]. However, such disposability does not mean the complete dissipation of polymeric waste. The rate of degradation of polymeric waste is dependent upon a number of factors, including the environmental, climatic and microbial characteristics of the region in which the waste is deposited [18–21]. Thus, the accumulation of plastic waste is one of the most pressing environmental challenges globally, affecting ecosystems and economies alike [22].

In the utilization of plastic waste, material recycling takes priority, followed by chemical recycling (pyrolysis) and biodegradation [23,24]. Biodegradation processes can be triggered by a variety of fungal and bacterial cultures, depending on the physicochemical characteristics of the plastic materials to be degraded. They also depend on environmental and physical factors, such as pH, temperature, the availability of nutrients, inducers, inhibitors, the source of carbon, the presence of the contaminant, etc. [25–29]. The molecular weight of the polymer is another decisive factor for the determination of the fate of polymers in the environment; i.e., polymers with a high molecular weight are less susceptible to degradability than those with a low molecular weight [30]. Temperature is a crucial parameter affecting biodegradation rates, as most studies indicate that microbial degradation takes place at a temperature of 20 °C or higher, while a recent study identified 34 cold-adapted microbial strains when examining the plastisphere of polymeric materials buried in alpine and arctic soils [31]. Biodegradability was found to be inversely proportional to molecular weight in persistent and biodegradable homopolymers in different organisms in both terrestrial and aquatic ecosystems [32–36]. Furthermore, the degradation mechanism is also dependent upon the type of organism being used for polymeric waste degradation.

The main obstacles that render the biotechnological recycling of plastic waste unsustainable are low efficiency and slow degradation. Consequently, an intensive search for microorganisms—often originating from environmental sources or insect digestive tracts—capable of decomposing polyethylene, polypropylene, polystyrene, polyethylene terephthalate, polyvinyl chloride, polytetrafluoroethylene, and polyurethanes (PURs) is

being undertaken [37–40]. However, some opinions suggest that the real utility of microbial decomposition is overemphasized [41]. PURs, a versatile group of polymers, significantly contribute to this issue due to their widespread use and resistance to degradation. Representing 7.7% of global plastic production, PURs are particularly resistant to biodegradation due to their varied chemical structures, composed of various isocyanates and polyols [42]. Recent studies, however, have made strides in investigating the biodegradation of PURs by microorganisms, fungi, and enzymes, offering promising insights into sustainable waste management solutions [43–50]. The susceptibility of plastics to degradation largely depends on their chemical structure. Nonetheless, some bacteria and fungi have demonstrated the ability to break down PURs under specific conditions. Several research studies have focused on the potential of microbial consortia to degrade synthetic polymers for use in bio-recycling processes [51]. Fungi, particularly from the *Ascomycota* phylum, have shown significant potential in degrading PURs. Fungi such as *Aspergillus*, *Cladosporium*, and *Penicillium* secrete specific enzymes such as hydrolases and ureases that contribute to the breakdown of plastic polymers [52,53]. The formation of biofilms by these fungi facilitates the degradation of PURs through the penetration of their enzymes into the material's structure. Soil microorganisms were found to be capable of mineralizing aliphatic PUR foams at a rate of 30–50% over six months, indicating that fungi and bacteria in soil environments can utilize PURs as both an energy and nutrient source, primarily carbon [25,26,28,29]. These findings underscore the significance of soil microorganisms in the natural degradation of synthetic polymers [54].

While significant progress has been made in understanding the microbial and enzymatic degradation of PURs, challenges remain. Microbial degradation is slower compared to physical and chemical methods; however, it holds great promise for long-term sustainability. Several types of enzymes are known to be involved in the degradation of polyurethanes, all of them belonging to the hydrolase (EC 3) enzyme class. Since polyurethanes are polymers containing carbamate links among their monomers, both ester hydrolases (EC 3.1) and non-peptide amidases (EC 3.5) can degrade them. Among these, the two most important types are carboxylic ester hydrolases (EC 3.1.1) and urea amidohydrolases (EC 3.5.1). Within the group of carboxylic ester hydrolases, carboxylesterases (EC 3.1.1.1) and lipases (EC 3.1.1.3) play significant roles. Interestingly, cutinases (EC 3.1.1.74) have increasingly been reported to hydrolyze polyurethanes. Enzymes termed polyesterases and polyurethanases degrade polyurethanes but are not classified within the enzyme nomenclature database [55,56]. Polyurethanases, such as the enzyme produced by *Pseudomonas chlororaphis*, directly degrade polyurethane molecules [22,29,51–53,57,58]. Polyesterases, like those produced by *Thermobifida fusca*, are specialized for hydrolyzing polyesters but can also degrade certain forms of polyurethanes [59,60]. Esterases, including enzymes produced by *Thermobifida alba*, and lipases, such as *Candida antarctica* lipase B, primarily contribute to the degradation of polyester-based polyurethanes [48,53,57,60]. Ureases, like those produced by *Bacillus megaterium*, can also break down the urethane moiety [51]. Cutinases, such as those produced by *Thermobifida cellulosilytica*, exhibit broader substrate specificity and are capable of degrading both polyesters and polyurethanes [61–64]. Enzymatic degradation of PURs has gained attention as a potential method for reducing plastic waste. An enzymatic pre-treatment strategy has been proposed, utilizing enzymes to promote surface oxidation and improve the hydrophilicity of plastics, thereby enhancing microbial attachment and accelerating biodegradation [22,27,65,66]. This approach offers an environmentally more friendly alternative to traditional chemical and physical treatments, which often result in secondary pollution and high energy consumption. Furthermore, enzymes such as polyester-degrading esterases and urethane bond-degrading amidases have been shown to play crucial roles in PUR degradation in *Streptomyces* species [58,67].

Proteomic analyses revealed that PUR degradation intermediates are incorporated into bacterial metabolic pathways, offering insights into how these microorganisms can be leveraged for industrial-scale plastic recycling [68]. The most effective solutions can be achieved by integrating various techniques. Combining microbial degradation with enzymatic pre-treatment or physical techniques can accelerate the process and enhance the overall efficiency of PUR waste treatment [21,52,57,69,70].

The pedospheric zone, which is a recipient of a number of pollutants and toxins, contains entities that can be effectively utilized for environmental detoxification [6,7,71–76]. The soil profile serves as a breeding ground for the growth and sustenance of various microorganisms with diverse metabolic pathways. These microbes can serve as biotechnological tools capable of degrading and transforming different types of toxins, including polymeric materials, by releasing enzymes that convert intact substances into byproducts [58,65,77–79]. The consumption of polymers by pedospheric microbes to meet their energy needs leads to structural variations in the polymers, such as reduction in molecular weight. These structural and morphological changes can be observed using ultraviolet spectroscopy or other standard testing methods. The initiation of polymer degradation is marked by the effective conversion of a large macromolecular unit into smaller repetitive units through the process of mineralization [25,26,29]. Microbe-driven transformation mechanisms for various contaminants, including polymeric degradation, can be divided into four general steps: adherence, assimilation, fragmentation and mineralization [43,80–82].

Currently, reported cases of environmental deterioration due to plastic pollution present a significant challenge for humanity, particularly given the remarkable persistence of polymers in the ecosystem [17]. Human responses to such environmental deterioration should prioritize completely avoiding the use of toxic products, including polymeric substances, and focus on identifying suitable green alternatives for plastics [27,83–85]. However, considering the widespread use of plastic commodities and their affordability, developing viable alternatives requires extensive research and capacity building. Under these circumstances, where plastic pollution has become an unavoidable issue, human efforts should center on the adoption of greener technologies for the prevention and minimization of plastic pollution. To date, numerous microbial species have been utilized for the degradation of polymeric substances [86,87].

The objective of this study was to isolate and identify polymer-degrading fungi and bacteria from a dumpsite in Rawalpindi and to develop a cost-effective framework for the microbial degradation of polymers. Additionally, the efficiency of fungal and bacterial species in degrading polymers was quantified through analysis. This research paves the way for further studies in this field, particularly in developing countries like Pakistan, where waste management is usually neglected.

2. Materials and Methods

2.1. Materials

The synthetic polymers selected for current investigation were low-density polyethylene (LDPE), represented by plastic shopping bags, and PUR, represented by plastic foam. A local dumpsite located in the vicinity of Rawalpindi, Pakistan (33.5651° N, 73.0169° E), was selected as a testing site for the isolation of fungal and bacterial cultures. The isolated microbial cultures were identified using both microscopic and macroscopic methods to examine a wide range of morphological and structural features. The microbial species derivation process involved several steps, including isolation, culturing, the identification of species, sample preparation, the addition of samples to a salt solution culture medium (a modified Czapek Dox liquid medium containing 2.0% $NaNO_3$, 0.7% KH_2PO_4, 0.3% K_2HPO_4, 0.5% KCl, 0.5% $MgSO_4.7H_2O$, 0.1% $FeSO_4.7H_2O$ and 0.05% NaN_3), inocula-

tion of the salt solution and, finally, incubation of the sample for one month. This process was followed by various analyses, including weight loss measurements, Fourier transform infrared spectroscopy (FT-IR), thermogravimetric analysis (TGA) and stereo and electron microscopic analysis. All of the chemicals used in this current research were of analytical grade, with 99.99% purity, and were utilized without further purification. Deionized water (DW) was used for all the experiments.

2.2. Pedospheric Sampling

A total of three sampling sites were selected for soil collection in Rawalpindi, Pakistan (Figure 1). Soil samples were collected at a depth of 10 cm, placed in zip-lock bags and transported to the laboratory for further analysis. After transportation, the soil samples from the selected dump sites were mixed in equal proportions to prepare a composite sample, which was then sieved. After the sieving process, three different soil dilutions were prepared by adding 1 g of soil from each site to 100 mL of DW for fungal and bacterial cultures coded as F1, F2 and F3 for fungal cultures and B1, B2 and B3 for bacterial cultures. For fungal cultures, the standard spread plate method was applied, while for bacterial cultures, the standard pour plate method was used.

(a)
(b)
(c)

Figure 1. Pedospheric microbial sampling, isolation and identification: (**a**) sampling site at a dumping place at Rawalpindi, Pakistan, showing plastic garbage; (**b**) fungal strains, i.e., *Aspergillus flavus*, *A. terreus*, *A. clavatus* and *A. niger*; and (**c**) bacterial strains, i.e., white, light-yellow, orange, pink and dark-yellow bacterial colonies.

2.3. Microbial Strain Selection and Micro/Macro Identification

The microbial cultures selected for polymer degradation included both fungal and bacterial cultures. All of the experiments were performed in triplicate, and average values were calculated. The initial fungal plates displayed colonies with varying growth patterns. A small sample from each colony was collected using a sterilized needle and inoculated onto freshly prepared plates containing isolates of the pure fungal colonies on potato dextrose agar (PDA). A total of four distinct fungal colonies were isolated, identified and used for polymeric degradation of LDPE and PUR, as shown in Figure 1. Fungal strains were identified using the microtiter plate procedure of Dobranic and Zak [88], employing the Biolog Microbial Identification System (Biolog, Inc., Hayward, CA, USA). This system examines the ability of the microbes to assimilate or oxidize substances from pre-selected panel carbon sources. The Biolog Microplate tests generated characteristic wells with turbidity changes, leading to the formation of a metabolic fingerprint. The panel, comprising 95 types of biochemical tests, facilitated fungal strain identification through a standardized micro-method. PU and LDPE degradation was analyzed using weight loss measurements, FT-IR using an FTIR-8400 spectrometer (Shimadzu Corp., Kyoto, Japan), stereomicroscopic analysis and TGA using a TGA 8000 thermogravimetric analyzer (PerkinElmer Inc., Shelton, CT, USA) based on methodology from the scientific literature [89]. Additionally, selected polymer samples were further examined using scanning electron microscopy (SEM) with a BK-EM6900 (KYKY Technology Co., Beijing, China) scanning electron microscope operated at 20 kV.

Bacterial isolates that exhibited clear zones of hydrolysis around their colonies, indicating PU degradation, were selected for further investigation. These isolates were identified through macroscopic characteristics (colony morphology, pigment, shape, size, margin, surface), microscopic features (Gram staining, shape, cell arrangement, granulation, spore presence, motility) and biochemical properties, following the standard determinative bacteriology guidelines [90]. Correct identification of the fungal strains in the current work was accomplished based on the scientific literature [88,89,91], via reading and comparison between the phenotypic fingerprint of each plate and the Biolog Microbial Identification Systems Database. Consequently, the identified fungal cultures were *Aspergillus flavus*, *A. terreus*, *A. clavatus* and *A. niger*.

In addition to fungal cultures, various bacterial colonies were obtained through the pour plate method using nutrient agar (NA). Five distinct bacterial colonies appeared on the Petri dishes, which were part of the triplicate sets. All the colonies exhibited considerable differentiation, as reflected in their respective morphological traits. A single isolate of each bacterial colony was collected using a sterile loop and streaked onto freshly prepared nutrient agar plates [89]. The five bacterial cultures used for polymeric degradation are shown in Table 1 and Figure 1.

Table 1. Genus of bacterial cultures and their codes.

Bacteria	Colony Color	Code
Coccus	Orange	C1
Coccus	Light yellow	C2
Coccus	Pink	C3
Filamentous	Dark yellow	F
Rod shaped bacteria	White	B

Macro-level identification of cultures was carried out based on their morphological characteristics, such as color, shape and size, as observed with the human naked eye, following methods in the scientific literature [88,89,91]. For instance, observations included the color and shape of fungal colonies, etc. Bacterial colonies were also distinctly visible to

the naked eye [91]. Micro-level identification included characterization with the help of a light microscope.

2.4. Inoculum Preparation

The inoculum density was standardized for both fungal and bacterial cultures to ensure uniform exposure to the polymeric materials (LDPE and PUR). Fungal and bacterial inoculums were prepared in PDA and NA, respectively. Each inoculum was prepared in conical flasks by adding a small cultured sample of the respective fungal or bacterial culture using a sterilized needle, into 100 mL corresponding broths. PDA broth inoculums were placed in a shaker at 120 rpm and 35 °C for 72 h [89]. Selected polymeric samples, LDPE and PUR, were then aseptically added to the solution [92]. For the experiment, 24 conical flasks for fungal cultures and 30 conical flasks for bacterial cultures were prepared, each containing 50 mL of the solution. Each flask was loaded with one sample of PUR and one sample of LDPE. These flasks were placed in a shaker at 120 rpm and 35 °C temperature for 48 h. After two days, the treated samples were removed from the shaker and analyzed. Inoculum homogeneity was ensured by thoroughly mixing the inoculated broth before distributing it into the conical flasks. Additionally, microbial growth in the broth was periodically assessed by sampling and confirming uniformity in colony-forming unit counts (in the range of 2–8 $\times$ 10^9 cfu/mL) through plating on nutrient agar. Optical density measurements at a wavelength of 600 nm were also conducted to verify consistent inoculum density across all flasks. This ensured uniform microbial exposure to the polymeric materials during the degradation process. The degradation of polymeric samples, LDPE and PUR, was studied using the salt solution listed in Table 1.

2.5. Sample Analysis Harvesting

After the incubation period, the growth of each species was clearly observed in each media bottle. Pre-weighed LDPE and PUR samples were added to 50 mL of a mineral salt solution, followed by the addition of 10 mL of inoculum into the conical flasks. These flasks, prepared for each species, were analyzed over the course of one month. Each flask was labeled with the specific fungal or bacterial species, corresponding to weekly sample intervals. These flasks were incubated for one month at 120 rpm and 35 °C. After the incubation period, the samples were harvested. The collected samples were washed first with distilled water and then with 70% ethanol before being dried under laminar airflow for 24 h [93].

2.6. Biodegradative Potential

To accurately estimate the activity of the selected pedospheric microbial cultures against LDPE and PUR, various analytical tools and measurements were employed. The weight loss of polymer was measured by first determining their initial weights before any microbial action on the samples. After one month of incubation, the dried samples were weighed again to assess the degradation rate using a comparison method [94]. The weight loss of selected polymers was calculated using Equation (1) [95].

$$\text{Weight loss} = \frac{W_1 - W_2}{W_1}\text{x}100 \tag{1}$$

where W_1 is the initial weight, and W_2 is the final weight of the polymer.

After the four-week incubation of PUR and LDPE films in a liquid medium, these samples were then subjected to the FT-IR to evaluate degradation based on the variations in the functional group through vibrational analysis. TGA was also conducted to perform thermal analysis of microbe-driven polymeric degradation. This method monitored the

physicochemical properties of the materials as a function of rising temperature (continuous heating rate) or time (continuous mass loss). The temperature range for the experiment was 25 °C to 600 °C for LPDE and PUR under an inert N_2 atmosphere. A stereomicroscope was used to study the surface of the solid specimens. Prepared samples were placed under a microscope to examine the surficial parameters, highlighting the occurrence and extent of biodegradation after microbial treatment.

3. Results and Discussion

Biodegradation is an eco-friendly, sustainable and cost-effective approach to plastic pollution prevention and minimization. This current work effectively utilized microbial biotechnological tools derived from the pedospheric zone of a dumpsite for the degradation of polymeric pollutants. To assess PUR and LDPE biodegradation by microbial strains, different testing methods, including weight loss measurements, FT-IR, TGA and stereo and electron microscopic analysis were employed [41]. The fungus *Aspergillus versicolor* isolated from soil was found to degrade up to 55% of PURs after one month of exposure [21]. Similarly, *Tenebrio molitor* larvae could degrade PUR foam by 35% in 17 days, assisted by gut microorganisms such as *Lactococcus* and *Pseudomonas* species [57,96]. Several studies have also demonstrated the potential of fungi to degrade PURs, such as *Cladosporium* sp. P7, which achieved a degradation rate of 94.5% of Impranil DLN-SD [52]. These findings suggest that microbial communities, both in soil and within insect digestive systems, play a crucial role in breaking down PUR materials [38,46,48,49,53,57,58,60,97–101]. Polymeric samples exhibited substantial degradation when exposed to microbial strains, as evidenced by weight reduction resulting from the adherence of microbial spores, which utilized polymeric substrate as a food and carbon source [102,103]. PUR samples treated with five bacterial and four fungal cultures showed varying rates of weight reduction. Weight reduction was initiated by the formation of a biofilm over a polymeric surface by bacterial and fungal cultures. An observable increase in the proliferation rate of the biofilm cells occurs with an increase during the incubation period [104–106], leading to partial degradation [107].

The most significant weight reductions were observed in samples treated with the coccus (C3) and *A. niger* species. In contrast to the cocci, other bacterial cultures showed little to no degradation. Among the coccus isolates, C2 exhibited the least percentage of weight reduction due to the slow formation of microcolonies, which subsequently affected the degradation rate. Prolonged incubation periods enhanced the likelihood and extent of degradation. The results of this current study align with or even surpass previous findings, such as the bacterial degradation of LDPE shopping bags, which achieved a 22.8% weight loss over 90 days of incubation [95]. In this current study, the LPDE samples treated with coccus (C3) achieved a 50% weight reduction, while polymer samples treated with *A. niger* showed a 40% weight reduction.

A. clavatus and *A. terreus* exhibited comparatively reduced weight loss compared to other strains. This limited degradative potential, reflected in the weight reduction rate, can be attributed to their inability to form a biofilm on the polymeric surface, thereby failing to trigger significant degradation. The hydrophobicity of the bacterial environment, in the absence of a carbon source, was favorable for microbial adhesion to the polymeric surface [107–109]. Polymeric samples treated with rod-shaped bacteria (B) exhibited the least weight reduction, amounting to only 10% of the initial weight. After isolating fungal and bacterial cultures from the dumpsite pedospheric region, they were identified based on macroscopic and microscopic characteristics. The results of the analysis confirmed the identification of the fungal strains *A. flavus*, *A. terreus*, *A. clavatus* and *A. niger* (Table 2), while the bacterial cultures were classified as rod-shaped bacteria, coccus and filamentous forms (Table 3).

Table 2. Microscopic and macro-based identification of fungal species.

Fungal Species	Microscopic Characteristic	Macro-Based Characteristic	Microscopic Image
Aspergillus niger	Spores appear to be dark brown or black in color.	The colony, dense and thick and black in color, grows within two or three days.	
Aspergillus flavus	Spores appear to be yellowish green to dark green. These spores are connected to large fungal hyphae.	The colony, light green to yellowish green in color, with a smooth surface, grows within a week.	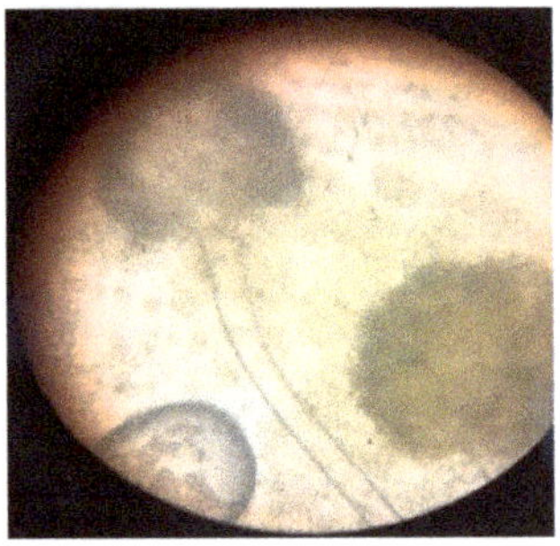
Aspergillus clavatus	Microscopic identification revealed the fungal body to be club shaped, having a spore color of medium brown to dark brown.	The colony, light green to olive green in color, grows within 2 days, and exudates are also visible on the surface.	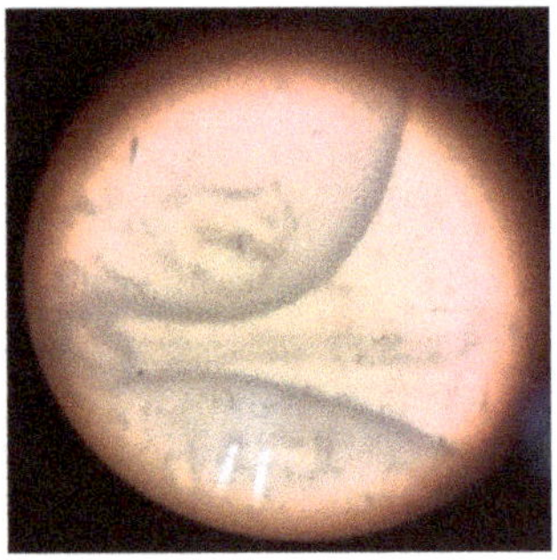
Aspergillus terreus	Identified by its short and round conidia attached to long hyphae, the color of these conidia is yellowish green to light green.	The colony, yellowish brown in color, grows very quickly within a few days.	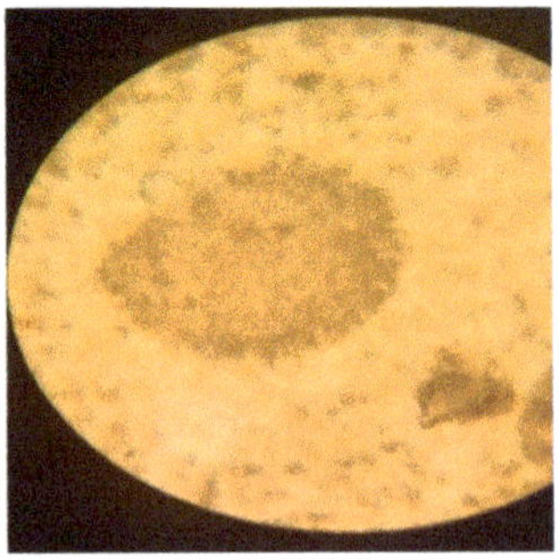

Table 3. Microscopic and macro-based identification of bacterial cultures.

Bacterial Colony Color	Stain Color	Shape	Characteristics	Microscopic Image
Light yellow	Purple	Round	Gram+, Coccus	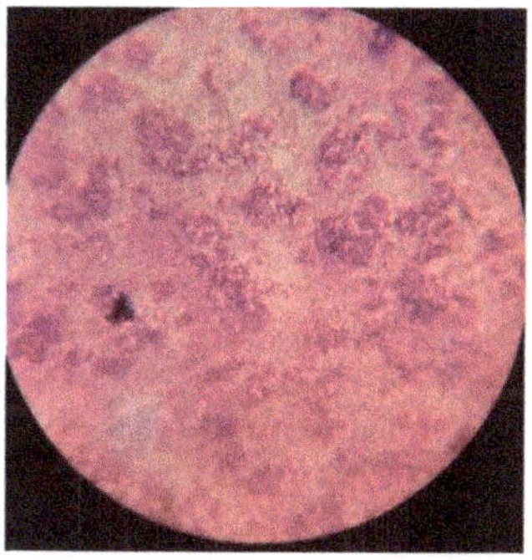
Orange	Purple	Round	Gram+, Coccus	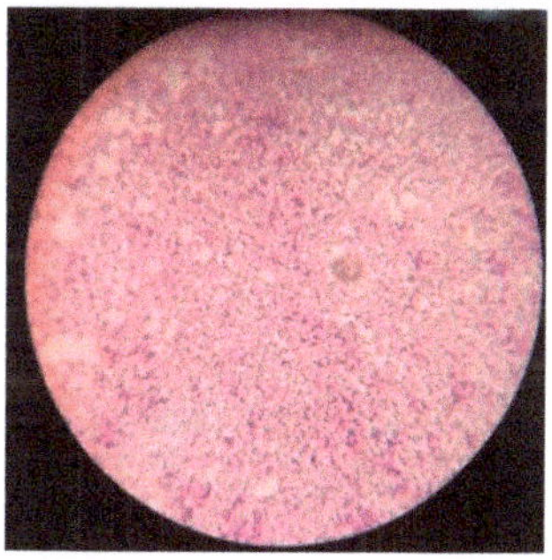
Pink	Purple	Round	Gram+, Coccus	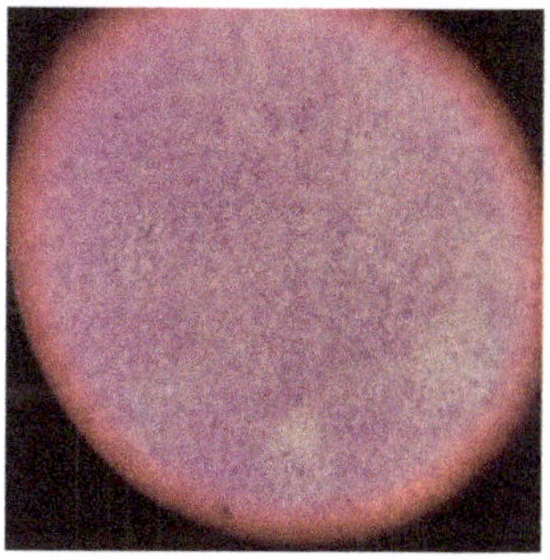
Dark yellow	Purple	Filamentous	Gram+, Filamentous	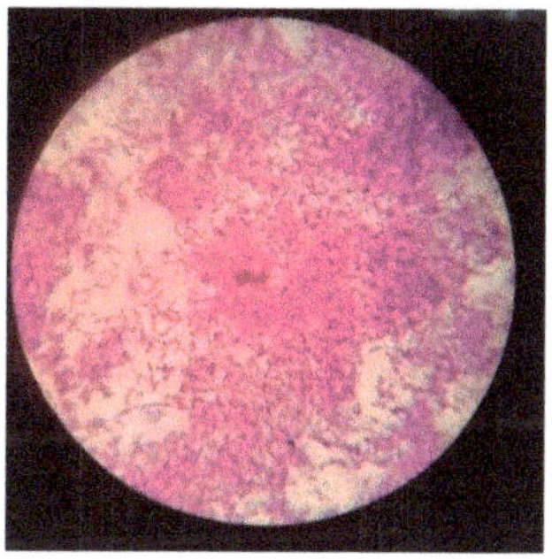

Table 3. *Cont.*

Bacterial Colony Color	Stain Color	Shape	Characteristics	Microscopic Image
White	Pink	Rod	Gram−, rod-shaped bacteria	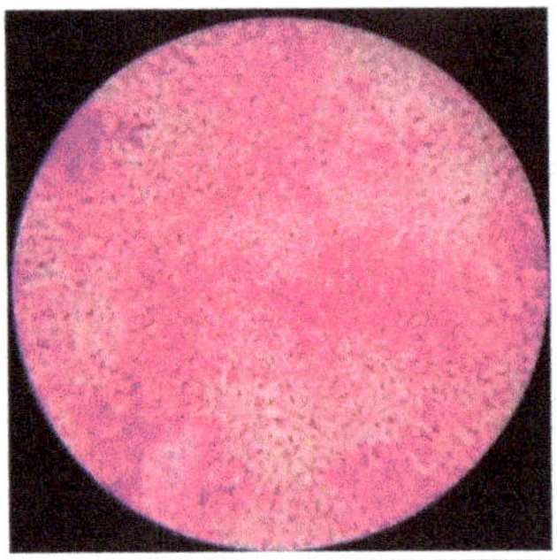

The weight of the polymeric samples was compared before and after the addition of the fungal or bacterial cultures. Throughout the procedure, the polymer samples were left undisturbed to ensure that the fungal and bacterial cultures utilized the polymers as their sole carbon source to meet their nutritional requirements. This utilization indirectly reflected the deterioration of the polymeric structure, leading to its conversion. This assay demonstrated the total reduction in polymeric weight over the experimental period. The final weight of the polymer samples expressed a profound reduction, indicating the potential of fungal and bacterial biotechnological tools in degrading the selected polymeric waste materials (Tables 4 and 5). Among fungal cultures, *A. niger* exhibited the highest weight reduction, while coccus bacterial cultures showed the most pronounced weight reduction in both polymeric samples. The weight reduction in LDPE samples treated with *A. niger* and coccus bacterial cultures was 40 and 50%, respectively. For PUR samples, the reduction was 55% with *A. niger* and 40% with the coccus bacterial culture.

Table 4. Final weight of the selected synthetic polymers, low-density polyethylene (initial weight: 0.010 g) and polyurethane (initial weight: 0.011 g), upon treatment with different fungal species.

Polymer	Type of Fungal Culture	Final Weight of Samples (g)
Polyethylene	*Aspergillus flavus*	0.007
	Aspergillus niger	0.006
	Aspergillus clavatus	0.009
	Aspergillus terreus	0.008
Polyurethane	*Aspergillus flavus*	0.008
	Aspergillus niger	0.005
	Aspergillus clavatus	0.006
	Aspergillus terreus	0.009

Table 5. Final weight of the selected synthetic polymers, low-density polyethylene and polyurethane, upon treatment with different bacterial strains.

Polymer	Bacterial Species	Colony Color	Final Weight of Samples (g)
Polyethylene	*Coccus* sp.	Orange	0.006
	Coccus sp.	Pink	0.005
	Coccus sp.	Light yellow	0.009
	Rod-shaped bacteria sp.	White	0.009
	Filamentous sp.	Dark yellow	0.008
Polyurethane	*Coccus* sp.	Orange	0.008
	Coccus sp.	Pink	0.007
	Coccus sp.	Light yellow	0.010
	Rod-shaped bacteria sp.	White	0.009
	Filamentous sp.	Dark yellow	0.010

FT-IR analysis was conducted to evaluate the breakdown products of the reference and pedospheric microbe-treated samples, as reflected by vibrational parameters shown in Figure 2. FT-IR spectra of the selected polymer LDPE and PUR exhibited a variety of peaks in response to microbial treatment. The controlled LDPE sample displayed numerous peaks, as shown in Figure 2a. Absorbance peaks formed at 2928 cm^{-1} and 2852 cm^{-1} indicated the presence of C-H bonds, signifying the presence of alkanes. An absorbance peak of variable strength at 1464 cm^{-1} showed the presence of a benzene ring, represented as C=C. Strong peaks were identified at 1091 cm^{-1} and 1369 cm^{-1}, indicating the formation of stretching of C-O bonds. The peak at 2659 cm^{-1} is clearly indicative of the presence of alcoholic OH groups. Other peaks at 798 and 719 cm^{-1} depict the presence of =C-H bending bonds. The FT-IR spectra of the LDPE sample inoculated with *A. niger* revealed changes in the peaks corresponding to C-H bonds (alkanes), transitioning from 2928 cm^{-1} to 2920 cm^{-1} and from 2853 cm^{-1} to 2850 cm^{-1} (Figure 2b). The peak associated with C-O bonds steadily occurred at 1369 cm^{-1}, and the peak corresponding to benzene rings also remained unaltered after the microbial treatment. FT-IR spectra of the LDPE sample incubated with coccus exhibited variability in the absorbance peaks (Figure 2c). The peak corresponding to C-H bonds at 2928 cm^{-1} shifted to 2914 cm^{-1}, indicating a weakening of the bond. A significant increase in benzene ring peak from 1464 cm^{-1} to 1471 cm^{-1} was notable. Furthermore, there was a transition in the peak associated with the strong stretch of C-H bonds, which shifted from 1092 cm^{-1} to 1082 cm^{-1}.

PUR polymeric samples were also analyzed using FT-IR to identify the functional groups present, as shown in Figure 3. The FT-IR spectra of reference PUR samples, which were not treated with any microbes, displayed various peaks (Figure 3a). An absorbance peak at 3281 cm^{-1} corresponded to the presence of a strong O-H bond. Peaks at 2968 cm^{-1}, 2926 cm^{-1} and 2862 cm^{-1} indicated the presence of alkane C-H stretches with strong intensity. The FT-IR spectra of PUR samples treated with *A. niger* (Figure 3b) showed significant conversion, including the shift in the O-H bond peak from 3277 cm^{-1} to 3273 cm^{-1}. While the peak for alkanes remained unaltered, a second peak stretched from 2862 cm^{-1} to 2864 cm^{-1}. The alkene peak changed from 1637 cm^{-1} to 1631 cm^{-1}, and the benzene peak transitioned from 1535 cm^{-1} to 1541 cm^{-1}. A notable shift occurred in the absorbance peak of the strong O-H bond from 3281 cm^{-1} to 3257 cm^{-1}. The aromatic C=C stretch absorbance peak shifted from 1541 cm^{-1} to 1535 cm^{-1}. Microbial degradation of urethane bonds has been assumed to occur through ester bond hydrolysis [91]. The results of this current work regarding the bacterial decomposition of PUR (Figure 3c) align with previous studies on polymeric biodegradation using different microbial strains [89,91,107,110]. The absorbance peak at 2968 cm^{-1} remained unchanged in all bacterial samples after incubation, indicating no alterations in the C-H bonds and no evidence of cleavage. However, the decrease in the aromatic group C=C stretch values from 1541 cm^{-1} to 1533 cm^{-1}, 1535 cm^{-1}, 1518 cm^{-1} and 1527 cm^{-1} can be attributed to hydrolysis mediated by microbial esterase enzymes. This process excludes the likelihood of urease and protease enzymes performing other biochemical actions [110]. The disruption of PUR was confirmed by ester bond hydrolysis [91]. The observed stretching in the aromatic compound serves as an indicator of PUR degradation.

The LDPE polymer degraded by fungal and bacterial cultures was analyzed for thermal parameters using TGA. The biodegradation of LDPE via *A. niger*, shown in Figure 4, indicates that the process occured as single-step degradation. Initially, the weight of the LDPE polymer sample remained stable as the temperature increased to 420 °C, where a peak was observed due to sudden weight loss of the sample. This was followed by a consequential reduction in weight up to 95% at 480 °C, after which stability in weight loss was observed. A comparative analysis of the untreated reference sample and the

A. niger-treated sample demonstrated the efficacy of the fungal biotechnological tool in degrading LDPE. In the untreated sample, the mass remained stable at 100% up to 430 °C, at which point 15% weight loss occurred. The *A. niger*-treated sample exhibited similar thermal behavior compared to the control LDPE sample, with only one peak observed at 425 °C. The degradation of the polymer began at 420 °C, and complete mass loss was observed at approximately 485 °C.

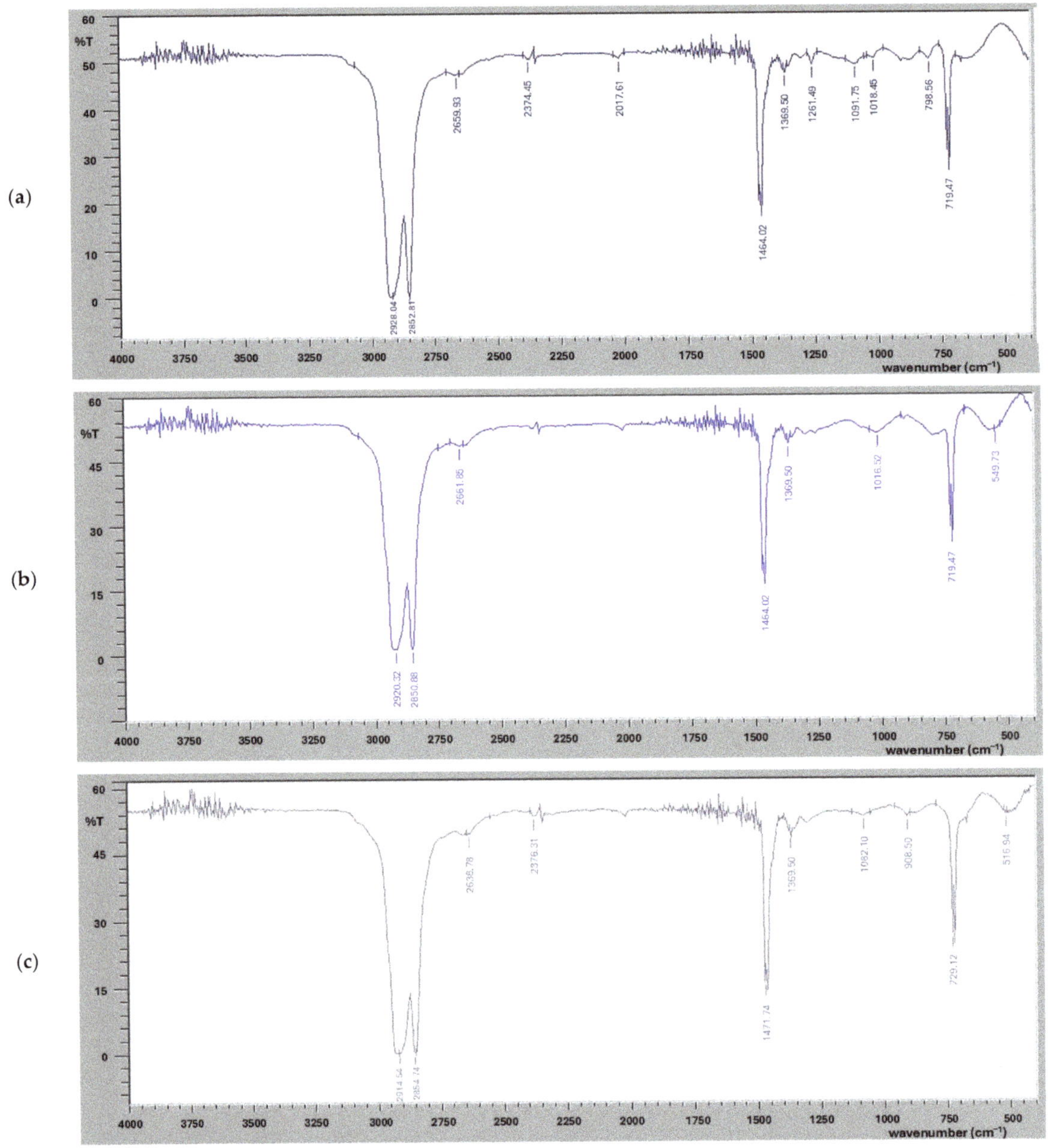

Figure 2. FT-IR spectra exhibiting the organic chemical profiles of the selected synthetic polymer LDPE, (**a**) control LDPE sample, (**b**) LDPE sample treated with fungal culture *Aspergillus niger* and (**c**) LDPE sample treated with bacterial strain coccus (C3).

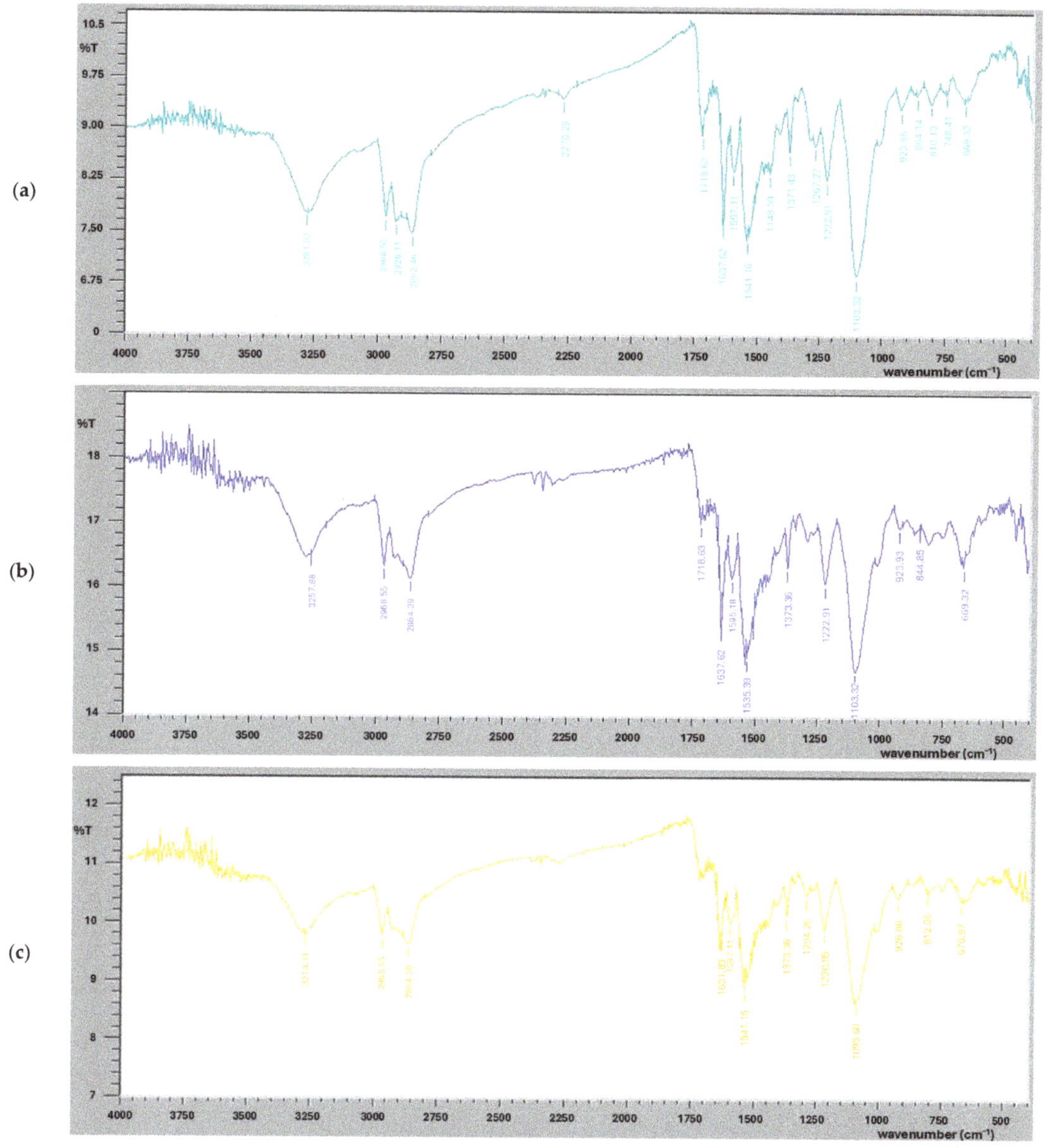

Figure 3. FT-IR spectra exhibiting the organic chemical profiles of the synthetic polymer PUR, (a) control PU sample, (b) PUR sample treated with fungal culture *Aspergillus niger* and (c) PUR sample treated with bacterial strain coccus (C3).

TGA was performed to understand the thermal behavior of the PUR polymer in the context of microbial degradation (Figure 5). The TGA results indicated that the degradation of PUR occurred as a two-step process. Initially, the polymer remained stable as the temperature increased. However, when it reached 260 °C, degradation began, resulting in a 58% weight reduction by 290 °C. A second peak was observed at 290 °C, marking the onset of further mass loss, which continued until 375 °C. At this point, the total mass lost was about 85%. Beyond this temperature, the weight loss stabilized despite further temperature increases. The *A. niger*-treated PUR sample exhibited two distinct peaks. The first peak

began at 250 °C and ended at 290 °C, corresponding to a weight loss of approximately 40%. The second peak started at 290 °C and ended at 380 °C, with a weight loss of about 90% (Figure 5b). The thermal behavior of the treated PUR polymer sample shifted from a two-stage to a one-stage degradation process. Degradation commenced at 40 °C, and the curve became linear up to 265 °C. Only one peak was observed at 265 °C. Notably, another peak present in the untreated PUR sample was absent in the treated sample. A maximum mass loss of 80% was observed at 375 °C.

(**a**)
(**b**)
(**c**)

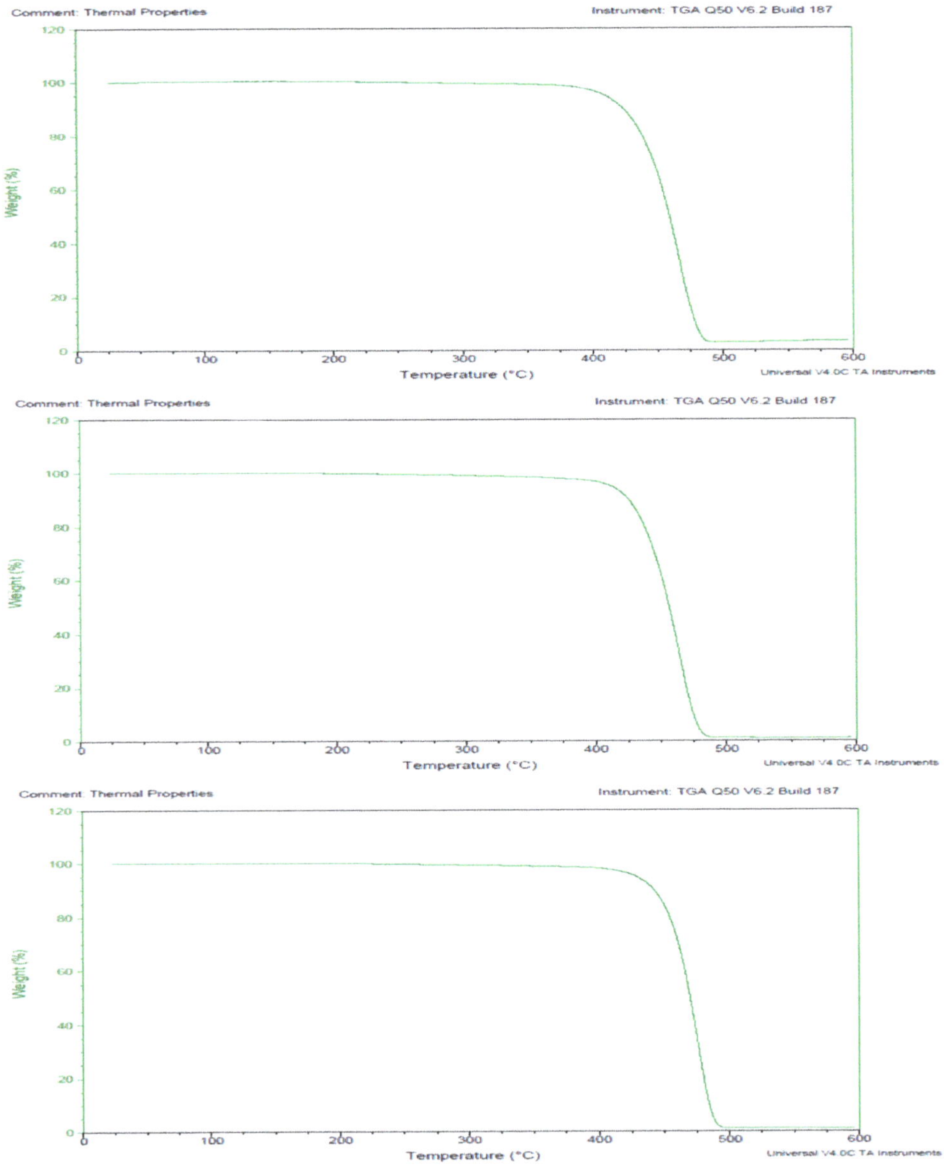

Figure 4. Thermograms exhibiting the breakdown pattern of the selected synthetic polymer LDPE, (**a**) thermogram exhibiting control LDPE sample, (**b**) thermogram exhibiting LDPE sample treated with fungal culture *Aspergillus niger* and (**c**) thermogram exhibiting LDPE sample treated with bacterial strain coccus (C3).

(**a**)

(**b**)

(**c**)

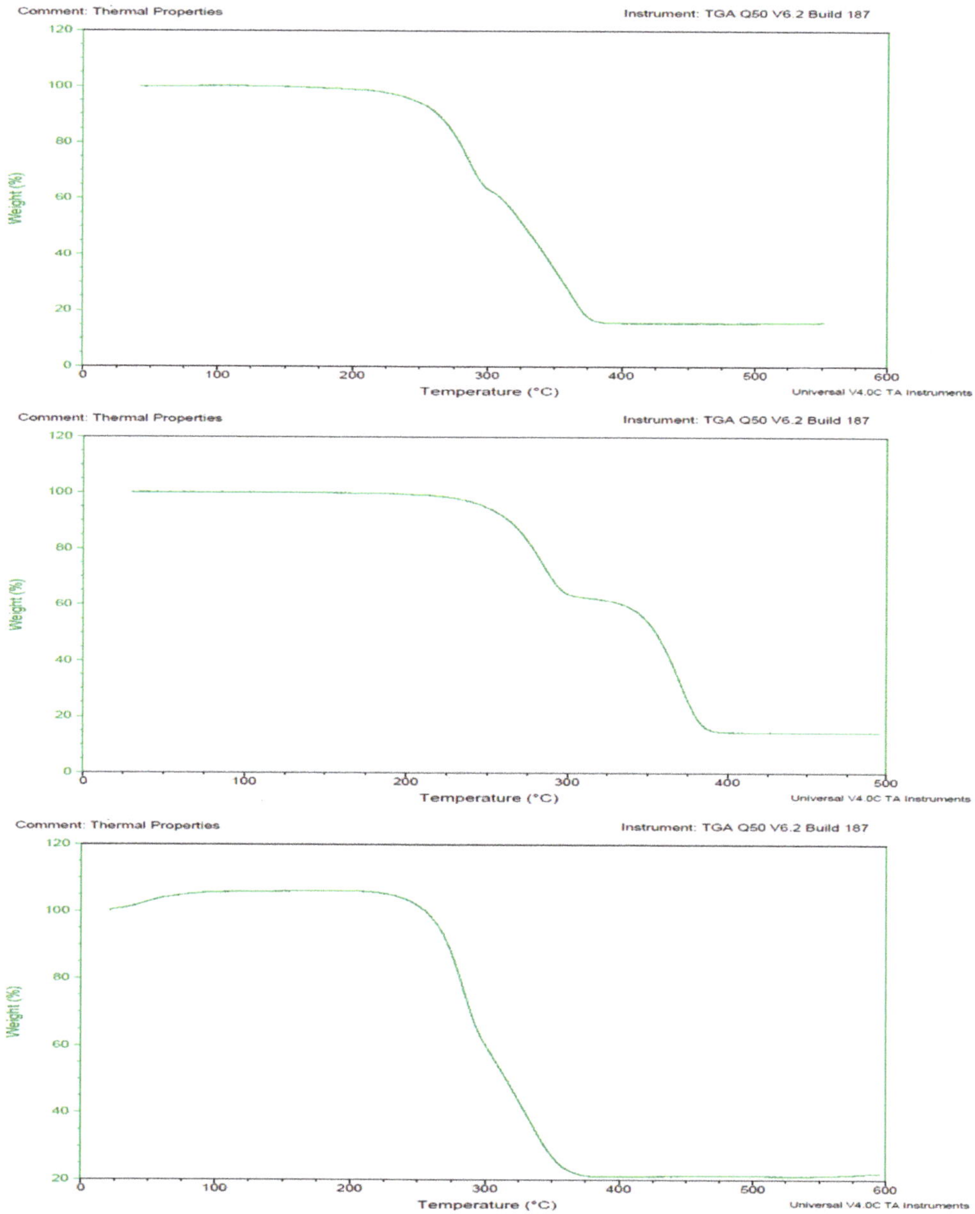

Figure 5. Thermograms exhibiting the breakdown pattern of the selected synthetic polymer PUR, (**a**) thermogram exhibiting control PU sample, (**b**) thermogram exhibiting PUR sample treated with fungal culture *Aspergillus niger* and (**c**) thermogram exhibiting PUR sample treated with bacterial strain coccus (C3).

In addition to the vibrational and thermal analyses, microbially degraded LPDE and PUR samples were also analyzed by stereomicroscopy and SEM for morphological variations in the treated samples. The sterile LDPE and PUR degraded samples were examined under a microscope to obtain a detailed three-dimensional view (Figure 6). Bio-eroded surfaces were identified as the primary cause of weight reduction [95]. Microscopic images

provided clear evidence of the successful deterioration of the polymeric substances following microbial treatment. The untreated LDPE sample exhibited a plain surface, whereas the untreated PUR sample (foam) displayed an irregular surface with roughness. Small, compactly bounded holes were also observed. After incubating the samples with fungal and bacterial cultures for four weeks, physical degradation was evident from stereomicroscopic images. The degradation manifested as pits, cracks and holes on the polymer surfaces, caused by the action of fungal and bacterial spores. The extent of degradation varied across microbial cultures, with some demonstrating significant deterioration. Initially, the controlled LDPE sample appeared very smooth, while the controlled PUR sample was compact. Post-incubation stereomicroscopic analysis revealed clear signs of degradation induced by bacterial cultures. For instance, the coccus-treated LDPE sample exhibited substantial surface changes, characterized by numerous small holes and dark spots, indicating surficial erosion.

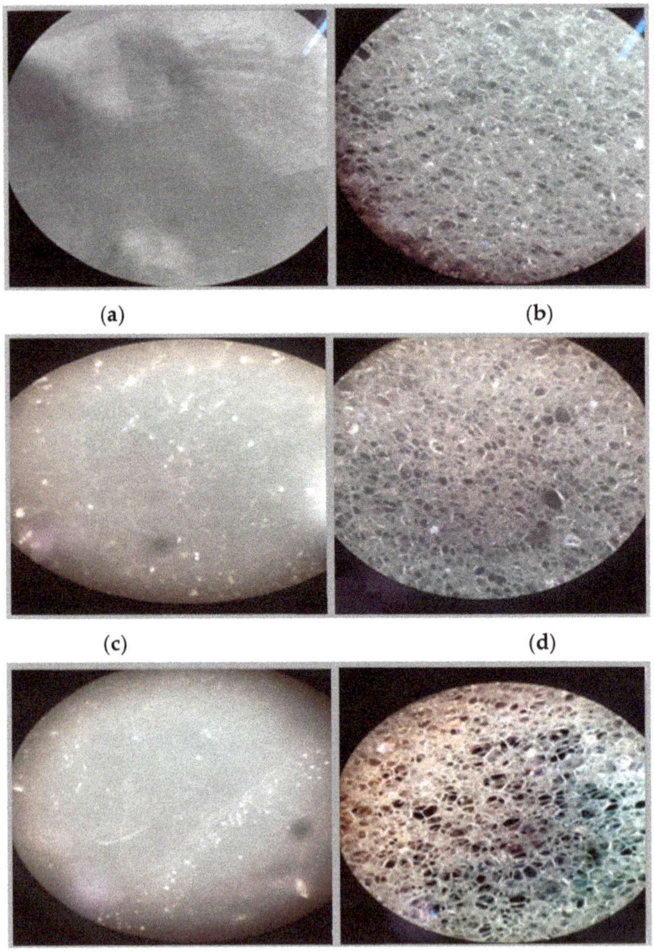

(a) (b) (c) (d) (e) (f)

Figure 6. Stereomicroscopic images of the selected synthetic polymers, (**a**) untreated LDPE, (**b**) untreated PUR, (**c**,**d**) *Aspergillus niger*-driven biodegradation in LDPE and PUR, respectively, and (**e**,**f**) coccus (C3)-driven biodegradation in LDPE and PUR, respectively.

SEM analysis revealed prominent degradation on the surface of polymer films, indicating the effective biodegradative capacity of the fungal strains used. This effect was particularly observed within LDPE films (Figure 7). It is assumed that the fungi caused structural changes in the films by secreting enzymes that break down the material and utilizing polyethylene as a carbon source for their growth. Fungal cultures have been reported to metabolize polyethylene, leading to weight reduction in the film [111]. In our hands, the *A. niger* stain demonstrated a weight reduction of 45% in a liquid medium and 9% in soil. The lower degradation efficiency in soil is attributed to the more favorable growth conditions in the liquid medium. These findings also highlight the importance of environmental factors in influencing the biodegradation of polyethylene films.

(**a**)

(**b**)

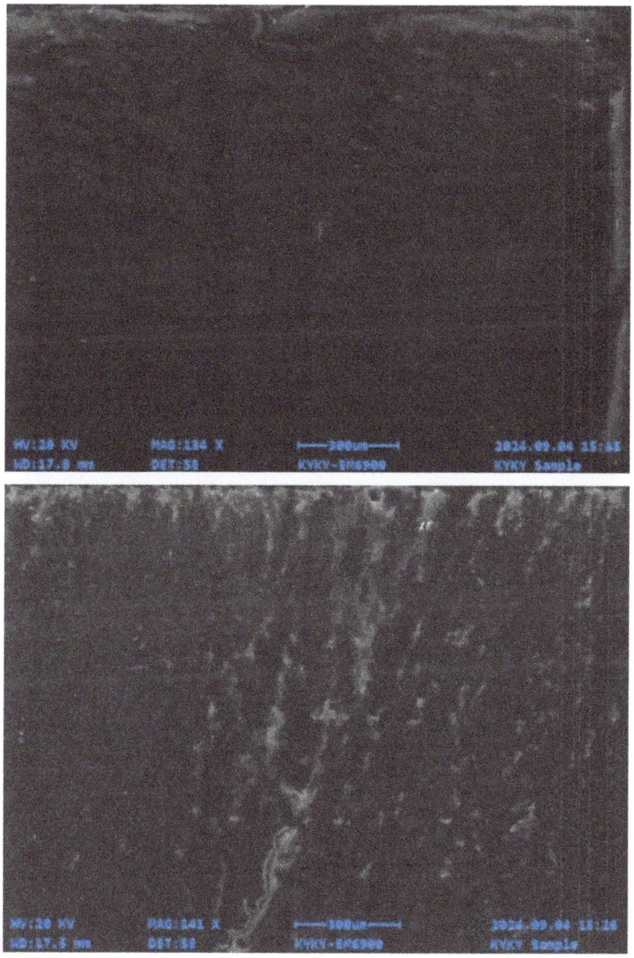

Figure 7. Scanning electron microscopic images of the degradation of LPDE by *Aspergillus niger* in liquid medium: (**a**) control and (**b**) *Aspergillus niger*-driven biodegradation in LDPE.

4. Conclusions

Plastics and other polymeric products pose a significant threat, not only to humans and animals but also to the entire ecosphere. The persistence and resistance of polymers to various degradation mechanisms exacerbate their impact on an already contaminated global environment. Among the diverse approaches to address this issue, the use of

microbial species as novel biotechnological tools for polymeric waste degradation emerges as a sustainable and cost-effective solution.

This research successfully utilized fungal and bacterial cultures isolated and identified from a local dumpsite, employing them as environmental remediators against LDPE and PUR polymers. Enhanced polymeric degradation driven by pedospheric microbes was achieved under controlled conditions, as reflected in FT-IR, TGA and stereomicroscopic analyses. Results showed that *A. niger* from fungal cultures and coccus among bacterial cultures contributed to the highest weight reduction in the polymeric samples. Weight reductions of 40% to 50% were observed for LDPE samples, and weight reductions of 40% to 55% were observed for PUR samples. Stereomicroscopic images clearly demonstrated surface morphological changes in the polymers before and after microbial treatment. The combined observations from stereomicroscopy, FT-IR spectral changes and TGA thermal degradation patterns strongly corroborate the occurrence of the biodegradation process. Notable, the FT-IR and TGA analyses also suggested potential alterations in the crystallinity of high-density polyethylene during microbial biodegradation, although further elucidation by SEM or atomic force microscopy is recommended. An additional important aspect that points toward further investigations related to the fungal strains isolated is that *Aspergillus* sp. are known for their ability to synthesize mycotoxins, such as aflatoxins in a strain-dependent manner, and are conditioned by environmental factors. As a part of the utility assessment of these strains to microbial decomposition of polymer waste materials, it must be confirmed that they do not produce aflatoxins. While this study primarily focused on the biodegradation potential of these microbes, the possible presence of aflatoxins was not assessed—a critical consideration for future research in order to ensure the safety and environmental sustainability of these strains in bioremediation. The promising results of this study highlight the practical viability of using pedospheric microbes as green and cost-effective solutions in polymeric waste remediation. Future research should explore optimizing these microbial strains to achieve higher degradation efficiencies. Additionally, the identification and evaluation of other pedospheric microbes for polymer waste disintegration may yield valuable insights and further advancements in this field.

Author Contributions: Conceptualization, S.I.; methodology, S.I. and A.S.; validation, S.I. and A.S.; formal analysis, S.I., M.O. and A.S.; investigation, M.N., S.J. and K.P.; writing—original draft preparation, M.N., S.J., S.I. and A.S.; writing—review and editing, K.P., M.O. and L.A.; visualization, M.N. and A.S.; supervision, S.I. and A.S.; funding acquisition, S.I., A.S. and L.A. All authors have read and agreed to the published version of the manuscript.

Funding: This work was funded by the National Research, Development and Innovation Office within the project "The feasibility of the circular economy during national defense activities" of the 2021 Thematic Excellence Program (grant No. TKP2021-NVA-22), led by the Centre for Circular Economy Analysis. Research funding was also provided by the Flagship Research Groups Program 2024 of the Hungarian University of Agriculture and Life Sciences.

Institutional Review Board Statement: Not applicable.

Data Availability Statement: The data presented in this study are contained within the manuscript.

Acknowledgments: The authors thank the Centre for Circular Economy Analysis and Knowledge founded by the Hungarian University of Agricultural and Life Sciences (MATE) for enrolling this international collaboration in its education profile.

Conflicts of Interest: Author László Aleksza was employed by the company Profikomp Environmental Technologies Inc. The remaining authors declare that this research was conducted in the absence of any commercial or financial relationships that could be construed as a potential conflict of interest.

References

1. Donohoe, M. Causes and health consequences of environmental degradation and social injustice. *Soc. Sci. Med.* **2003**, *56*, 573–587. [CrossRef] [PubMed]
2. Yeganeh, K.H. A typology of sources, manifestations, and implications of environmental degradation. *Manag. Environ. Qual.* **2020**, *31*, 765–783. [CrossRef]
3. Singh Jadaun, J.; Bansal, S.; Sonthalia, A.; Rai, A.K.; Singh, S.P. Biodegradation of plastics for sustainable environment. *Bioresour. Technol.* **2022**, *347*, 126697. [CrossRef] [PubMed]
4. Ridley, R.S.; Conrad, R.E.; Lindner, B.G.; Woo, S.; Konstantinidis, K.T. Potential routes of plastics biotransformation involving novel plastizymes revealed by global multi-omic analysis of plastic associated microbes. *Sci. Rep.* **2024**, *14*, 8798. [CrossRef] [PubMed]
5. Pandey, A.; Tripathi, P.H.; Pandey, S.C.; Pathak, V.M.; Nailwal, T.K. Removal of toxic pollutants from soil using microbial biotechnology. In *Microbial Biotechnology in Environmental Monitoring and Cleanup*; Pankaj, A.S., Ed.; IGI Global: Hershey, PA, USA, 2018; pp. 86–105. [CrossRef]
6. Jaffri, S.B.; Ahmad, K.S.; ul Saba, M.; Sher, G.; Sharif, M. Sorptive interactions evaluation of benomyl metabolites mecarzole with the varyingly selected minerals. *J. Int. Econ. Environ. Geol.* **2019**, *10*, 50–54. [CrossRef]
7. Ali, M.; Jaffri, S.B.; Ahmad, K.S.; Iqbal, S. Sorptive interactions of fungicidal 2-(4′-thiazolyl)benzimidazole with soils of divergent physicochemical composition. *J. Int. Econ. Envir. Geol.* **2019**, *10*, 97–104. [CrossRef]
8. Jaffri, S.B.; Ahmad, K.S. Interfacial engineering revolutionizers: Perovskite nanocrystals and quantum dots accentuated performance enhancement in perovskite solar cells. *Critic. Rev. Solid State Mat. Sci.* **2020**, *46*, 251–279. [CrossRef]
9. Maqsood, Q.; Sumrin, A.; Waseem, R.; Hussain, M.; Imtiaz, M.; Hussain, N. Bioengineered microbial strains for detoxification of toxic environmental pollutants. *Environ. Res.* **2023**, *227*, 115665. [CrossRef]
10. Chettri, D.; Verma, A.K.; Verma, A.K. Bioaugmentation: An approach to biological treatment of pollutants. *Biodegradation* **2024**, *35*, 117–135. [CrossRef]
11. MacLeod, M.; Arp, H.P.H.; Tekman, M.B.; Jahnke, A.J.S. The global threat from plastic pollution. *Science* **2021**, *373*, 61–65. [CrossRef]
12. Bucci, K.; Tulio, M.; Rochman, C.J.E.A. What is known and unknown about the effects of plastic pollution: A meta-analysis and systematic review. *Ecol. Appl.* **2020**, *30*, e02044. [CrossRef] [PubMed]
13. Milligan, G.; Ward, R.J.; Marsango, S. GPCR homo-oligomerization. *Curr. Opin. Cell Biol.* **2019**, *57*, 40–47. [CrossRef] [PubMed]
14. Cárdenas Espinosa, M.J.; Colina Blanco, A.; Heipieper, H.J.; Eberlein, C. Screening and cultivating microbial strains able to grow on building blocks of polyurethane. In *Methods in Enzymology*; Elsevier: Amsterdam, The Netherlands, 2021; pp. 423–434. [CrossRef]
15. Mendes, J.F.; Martins, J.T.; Manrich, A.; Neto, A.S.; Pinheiro, A.C.M.; Mattoso, L.H.C.; Martins, M.A. Development and physical-chemical properties of pectin film reinforced with spent coffee grounds by continuous casting. *Carb. Polym.* **2019**, *210*, 92–99. [CrossRef] [PubMed]
16. Tamayo, L.; Palza, H.; Bejarano, J.; Zapata, P.A. Polymer composites with metal nanoparticles: Synthesis, properties, and applications. In *Polymer Composites with Functionalized Nanoparticles*; Pielichowski, K., Majkapp, T.M., Eds.; Elsevier: Amsterdam, The Netherlands, 2019; pp. 249–286. [CrossRef]
17. North, E.J.; Halden, R.U. Plastics and environmental health: The road ahead. *Rev. Environ. Health* **2013**, *28*, 1–8. [CrossRef]
18. Haider, T.P.; Völker, C.; Kramm, J.; Landfester, K.; Wurm, F.R. Plastics of the future? The impact of biodegradable polymers on the environment and on society. *Angewandte Chem. Int. Ed.* **2019**, *58*, 50–62. [CrossRef]
19. Scaffaro, R.; Maio, A.; Sutera, F.; Gulino, E.F.; Morreale, M. Degradation and recycling of films based on biodegradable polymers: A short review. *Polymers* **2019**, *11*, 651. [CrossRef]
20. Rajan, A.; Vijayalakshmi, S. New insights on polyurethane biodegradation. *AIP Conf. Proc. Int. Conf. Appl. Sci. Eng.* **2023**, *2690*, 020060. [CrossRef]
21. Rajan, A.; Ameen, F.; Jambulingam, R.; Shankar, V. Biodegradation of polyurethane by fungi isolated from industrial wastewater—A sustainable approach to plastic waste management. *Polymers* **2024**, *16*, 1411. [CrossRef]
22. Lee, S.; Lee, Y.R.; Kim, S.J.; Lee, J.S.; Min, K. Recent advances and challenges in the biotechnological upcycling of plastic wastes for constructing a circular bioeconomy. *Chem. Eng. J.* **2023**, *454*, 140470. [CrossRef]
23. Webb, H.K.; Arnott, J.; Crawford, R.J.; Ivanova, E.P. Plastic degradation and its environmental implications with special reference to poly(ethylene terephthalate). *Polymers* **2013**, *5*, 1–18. [CrossRef]
24. Aleksza, L. (Ed.) *Waste Management*; Profikomp Kft.: Gödöllő, Hungary, 2014. Available online: http://www.nyf.hu/agrtud/sites/www.nye.hu.agrtud/files/konyvek/SZIE_Konyv_Angol_teljes.pdf (accessed on 15 December 2024).
25. Zachinyaev, Y.V.; Miroshnichenko, I.I.; Zachinyaeva, A.V. Microbial degradation of polyurethane. *Russian J. Appl. Chem.* **2009**, *82*, 1321–1323. [CrossRef]

26. Gómez, E.F.; Luo, X.; Li, C.; Michel, F.C., Jr.; Li, Y. Biodegradability of crude glycerol-based polyurethane foams during composting, anaerobic digestion and soil incubation. *Polymer Degrad. Stab.* **2014**, *102*, 195–203. [CrossRef]
27. Devi, R.S.; Kannan, V.R.; Natarajan, K.; Nivas, D.; Kannan, K.; Chandru, S.; Antony, A.R. The role of microbes in plastic degradation. In *Environmental Waste Management*; Chandra, R., Ed.; CRC Press: Boca Raton, FL, USA, 2016; pp. 341–370. [CrossRef]
28. Gunawan, N.R.; Tessman, M.; Schreiman, A.C.; Simkovsky, R.; Samoylov, A.A.; Neelakantan, N.K.; Bemis, T.A.; Burkart, M.D.; Pomeroy, R.S.; Mayfield, S.P. Rapid biodegradation of renewable polyurethane foams with identification of associated microorganisms and decomposition products. *Bioresour. Technol. Rep.* **2020**, *11*, 100513. [CrossRef]
29. Skleničková, K.; Suchopárová, E.; Abbrent, S.; Pokorný, V.; Kočková, O.; Nevoralová, M.; Cajthaml, T.; Strejček, M.; Uhlík, O.; Halecký, M.; et al. Biodegradation of aliphatic polyurethane foams in soil: Influence of amide linkages and supramolecular structure. *Sci. Total Environ.* **2024**, *912*, 169062. [CrossRef]
30. Li, J.; Zhao, Y.; Jiang, Z.; Qiu, Z. Effect of low molecular weight poly(diethylene glycol adipate) on the crystallization behavior and mechanical properties of biodegradable poly(L-lactide) in their partially miscible blends. *Polymer Degrad. Stab.* **2019**, *159*, 217–223. [CrossRef]
31. Rüthi, J.; Cerri, M.; Brunner, I.; Stierli, B.; Sander, M.; Frey, B. Discovery of plastic-degrading microbial strains isolated from the alpine and Arctic terrestrial plastisphere. *Front. Microbiol.* **2023**, *14*, 1178474. [CrossRef]
32. Hadad, D.; Geresh, S.; Sivan, A. Biodegradation of polyethylene by the thermophilic bacterium *Brevibacillus borstelensis*. *J. Appl. Microbiol.* **2005**, *98*, 1093–1100. [CrossRef]
33. Kale, G.; Auras, R.; Singh, S.P.; Narayan, R. Biodegradability of polylactide bottles in realand simulated composting conditions. *Polym. Test.* **2007**, *26*, 1049–1061. [CrossRef]
34. Chinaglia, S.; Tosin, M.; Degli-Innocenti, F. Biodegradation rate of biodegradable plastics at molecular level. *Polym. Degrad. Stabil.* **2018**, *147*, 237–244. [CrossRef]
35. He, L.; Ding, J.; Yang, S.S.; Zang, Y.-N.; Pang, J.-W.; Xing, D.; Zhang, L.-Y.; Ren, N.; Wu, W.M. Molecular-weight-dependent degradation of plastics: Deciphering host-microbiome synergy biodegradation of high-purity polypropylene microplastics by mealworms. *Environ. Sci. Technol.* **2024**, *58*, 6647–6658. [CrossRef]
36. Seok, J.H.; Iwata, T. Effects of molecular weight on the marine biodegradability of poly(l-lactic acid). *Biomacromolecules* **2024**, *25*, 4420–4427. [CrossRef] [PubMed]
37. Ru, J.; Huo, Y.; Yang, Y. Microbial degradation and valorization of plastic wastes. *Front. Microbiol.* **2020**, *11*, 442. [CrossRef] [PubMed]
38. Chigwada, A.D.; Tekere, M. The plastic and microplastic waste menace and bacterial biodegradation for sustainable environmental clean-up a review. *Environ. Res.* **2023**, *231*, 116110. [CrossRef] [PubMed]
39. Yang, X.-G.; Wen1, P.P.; Yang, Y.F.; Jia, P.P.; Li, W.G.; Pei, D.S. Plastic biodegradation by *in vitro* environmental microorganisms and *in vivo* gut microorganisms of insects. *Front. Microbiol.* **2023**, *13*, 1001750. [CrossRef]
40. Cao, Y.; Bian, J.; Han, Y.; Liu, J.; Ma, Y.; Feng, W.; Deng, Y.; Yu, Y. Progress and prospects of microplastic biodegradation processes and mechanisms: A bibliometric analysis. *Toxics* **2024**, *12*, 463. [CrossRef]
41. Lear, G.; Maday, S.D.M.; Gambarini, V.; Northcott, G.; Abbel, R.; Kingsbury, J.M.; Weaver, L.; Wallbank, J.A.; Pantos, O. Microbial abilities to degrade global environmental plastic polymer waste are overstated. *Environ. Res. Lett.* **2022**, *17*, 043002. [CrossRef]
42. Sánchez-Reyes, A.; Gaytán, I.; Pulido-García, J.; Bureloa, M.; Vargas-Suárez, M.; Cruz-Gómez, M.J.; Loza-Tavera, H. Genetic basis for the biodegradation of a polyether-polyurethane-acrylic copolymer by a landfill microbial community inferred by metagenomics deconvolution analysis. *Sci. Total Environ.* **2023**, *881*, 163367. [CrossRef]
43. Ali, S.S.; Elsamahy, T.; Koutra, E.; Kornaros, M.; El-Sheekh, M.; Abdelkarim, E.A.; Zhu, D.; Sun, J. Degradation of conventional plastic wastes in the environment: A review on current status of knowledge and future perspectives of disposal. *Sci. Total Environ.* **2021**, *771*, 144719. [CrossRef]
44. Andler, R.; Tiso, T.; Blank, L.; Andreeßen, C.; Zampolli, J.; D'Afonseca, V.; Guajardo, C.; Díaz-Barrera, A. Current progress on the biodegradation of synthetic plastics: From fundamentals to biotechnological applications. *Rev. Environ. Sci. Bio/Technol.* **2022**, *21*, 829–850. [CrossRef]
45. Zampolli, J.; Mangiagalli, M.; Vezzini, D.; Lasagni, M.; Ami, D.; Natalello, A.; Arrigoni, F.; Bertini, L.; Lotti, M.; Di Gennaro, P. Oxidative degradation of polyethylene by two novel laccase-like multicopper oxidases from *Rhodococcus opacus* R7. *Environ. Technol. Innov.* **2023**, *32*, 103273. [CrossRef]
46. Maestri, C.; Plancher, L.; Duthoit, A.; Hébert, R.L.; Di Martino, P. Fungal biodegradation of polyurethanes. *J. Fungi* **2023**, *9*, 760. [CrossRef] [PubMed]
47. Taxeidis, G.; Nikolaivits, E.; Siaperas, R.; Gkountela, C.; Vouyiouka, S.; Pantelic, B.; Nikodinovic-Runic, J.; Topakas, E. Triggering and identifying the polyurethane and polyethylene-degrading machinery of filamentous fungi secretomes. *Environ. Pollut.* **2023**, *325*, 121460. [CrossRef] [PubMed]

48. Devi, R.J.; Usha, R. Microbial approaches for the plastic bioremediation and ecofriendly environmental sustainability. *Asian J. Chem.* **2023**, *35*, 289–300. [CrossRef]
49. Chow, J.; Perez-Garcia, P.; Dierkes, R.; Streit, W.R. Microbial enzymes will offer limited solutions to the global plastic pollution crisis. *Microb. Biotechnol.* **2023**, *16*, 195–217. [CrossRef]
50. Ibrahim, S.S.; Ionescu, D.; Grossart, H.-P. Tapping into fungal potential: Biodegradation of plastic and rubber by potent fungi. *Sci. Total Environ.* **2024**, *934*, 173188. [CrossRef]
51. Su, T.; Zhang, T.; Liu, P.; Bian, J.; Zheng, Y.; Yuan, Y.; Li, Q.; Liang, Q.; Qi, Q. Biodegradation of polyurethane by the microbial consortia enriched from landfill. *Appl. Microbiol. Biotechnol.* **2023**, *107*, 1983–1995. [CrossRef]
52. Liu, J.; Zeng, O.; Lei, H.; Xin, K.; Xu, A.; Wei, R.; Li, D.; Zhou, J.; Dong, W.; Jiang, M. Biodegradation of polyester polyurethane by *Cladosporium* sp. P7: Evaluating its degradation capacity and metabolic pathways. *J. Hazard. Mater.* **2023**, *448*, 130776. [CrossRef]
53. Khatura, S.; Simal-Gandara, J.; Acharya, K. Myco-remediation of plastic pollution: Current knowledge and future prospects. *Biodegr.* **2024**, *35*, 249–279. [CrossRef]
54. Struszczyk-Świta, K.; Drożdżyński, P.; Murawska, K.; Marchut-Mikołajczyk, O. PUF-immobilized *Bjerkandera adusta* dsm 3375 as a tool for bioremediation of creosote oil contaminated soil. *Int. J. Mol. Sci.* **2022**, *23*, 12441. [CrossRef]
55. IUBMB (International Union of Biochemistry and Molecular Biology). *Enzyme Nomenclature 1992*; Academic Press: San Diego, CA, USA, 1992; ISBN 0-12-227164-5.
56. Purich, D.L.; Allison, R.D. *The Enzyme Reference. A Connprehensive Guidebook to Enzynne Nonnenclature, Reactions, and Methods*; Academic Press: San Diego, CA, USA, 2002; ISBN 0-12-568041-4.
57. Orts, J.M.; Parrado, J.; Pascual, J.A.; Orts, A.; Cuartero, J.; Tejada, M.; Ros, M. Polyurethane foam residue biodegradation through the *Tenebrio molitor* digestive tract: Microbial communities and enzymatic activity. *Polymers* **2023**, *15*, 204. [CrossRef]
58. Pantelic, B.; Siaperas, R.; Budin, C.; de Boer, T.; Topakas, E.; Nikodinovic-Runic, J. Proteomic examination of polyester-polyurethane degradation by *Streptomyces* sp. PU10: Diverting polyurethane intermediates to secondary metabolite production. *Microb. Biotechnol.* **2024**, *17*, e14445. [CrossRef] [PubMed]
59. Jin, X.; Dong, J.; Guo, X.; Ding, M.; Bao, R.; Luo, Y. Current advances in polyurethane biodegradation. *Polymer Int.* **2022**, *71*, 1384–1392. [CrossRef]
60. Raczyńska, A.; Góra, A.; André, I. An overview on polyurethane-degrading enzymes. *Biotechnol Adv.* **2024**, *77*, 108439. [CrossRef] [PubMed]
61. Schmidt, J.; Wei, R.; Oeser, T.; Dedavid, E.S.L.A.; Breite, D.; Schulze, A.; Zimmermann, W. Degradation of polyester polyurethane by bacterial polyester hydrolases. *Polymers* **2017**, *9*, 65. [CrossRef]
62. Di Bisceglie, F.; Quartinello, F.; Vielnascher, R.; Guebitz, G.H.; Pellis, A. Cutinase-catalyzed polyester-polyurethane degradation: Elucidation of the hydrolysis mechanism. *Polymers* **2022**, *14*, 11. [CrossRef]
63. Jiang, Z.; Chen, X.; Xue, H.; Li, Z.; Lei, J.; Yu, M.; Yan, X.; Cao, H.; Zhou, J.; Liu, J.; et al. Novel polyurethane-degrading cutinase BaCut1 from *Blastobotrys* sp. G-9 with potential role in plastic bio-recycling. *J. Hazard. Mat.* **2024**, *472*, 134493. [CrossRef]
64. Romano, A.; Rosato, A.; Sisti, L.; Zanaroli, G.; Asadauskas, S.J.; Nemaniutė, P.; Bražinskienė, D.; Grigucevičienė, A.; Totaro, G. Enzyme-catalyzed polyurethane adhesive degradation. *React. Chem. Eng.* **2024**, *9*, 31–33. [CrossRef]
65. Ray, A.S.; Rajasekaran, M.; Uddin, M.; Kandasamy, R. Laccase driven biocatalytic oxidation to reduce polymeric surface hydrophobicity: An effective pre-treatment strategy to enhance biofilm mediated degradation of polyethylene and polycarbonate plastics. *Sci. Total Environ.* **2023**, *904*, 166721. [CrossRef]
66. Orlando, M.; Molla, G.; Castellani, P.; Pirillo, V.; Torretta, V.; Ferronato, N. Microbial enzyme biotechnology to reach plastic waste circularity: Current status, problems and perspectives. *Int. J. Mol. Sci.* **2023**, *24*, 3877. [CrossRef]
67. Tian, H.; Du, Y.; Luo, X.; Dong, J.; Chen, S.; Hu, X.; Zhang, M.; Liu, Z.; Abolfathi, S. Understanding visible light and microbe-driven degradation mechanisms of polyurethane plastics: Pathways, property changes, and product analysis. *Water Res.* **2024**, *259*, 121856. [CrossRef]
68. Wei, R.; Zimmermann, W. Microbial enzymes for the recycling of recalcitrant petroleum-based plastics: How far are we? *Microb. Biotechnol.* **2017**, *10*, 1308–1322. [CrossRef] [PubMed]
69. Heris, Y.S. Bacterial biodegradation of synthetic plastics: A review. *Bull. Natl. Res. Cent.* **2024**, *48*, 87. [CrossRef]
70. Thakur, B.; Singh, J.; Singh, J.; Angmo, D.; Vig, A.P. Biodegradation of different types of microplastics: Molecular mechanism and degradation efficiency. *Sci. Total Environ.* **2023**, *877*, 162912. [CrossRef] [PubMed]
71. Shaheen, I.; Ahmad, K.S.; Jaffri, S.B.; Zahra, T.; Azhar, S. Evaluating the adsorption and desorption behavior of triasulfuron as a function of soil physico-chemical characteristics. *Soil Environ.* **2016**, *35*, 99–105.
72. Zahra, T.; Ahmad, K.S.; Shaheen, I.; Azhar, S.; Jaffri, S.B. Determining the adsorption and desorption behavior of thiabendazole fungicide for five different agricultural soils. *Soil Environ.* **2017**, *36*, 13–19. [CrossRef]
73. Jaffri, S.B.; Ahmad, K.S. Augmented photocatalytic, antibacterial and antifungal activity of prunosynthetic silver nanoparticles. *Artif. Cells Nanomed. Biotech.* **2017**, *46* (Suppl. 1), 127–137. [CrossRef]

74. Jaffri, S.B.; Ahmad, K.S. *Prunus cerasifera* Ehrh. fabricated ZnO nano falcates and its photocatalytic and dose dependent in vitro bio-activity. *Open Chem.* **2018**, *16*, 141–154. [CrossRef]
75. Amjad, I.; Javaid, M.; Ikhlaq, K.; Gul, S.; Jaffri, S.B.; Ahmad, K.S. Adsorption-desorption mechanism of synthesized benzimidazole based fungicide 2-(3′-pyridyl) on selected soil minerals. *J. Int. Econ. Envir. Geol.* **2019**, *10*, 38–44. [CrossRef]
76. Jaffri, S.B.; Ahmad, K.S. Biomimetic detoxifier *Prunus cerasifera* Ehrh. silver nanoparticles: Innate green bullets for morbific pathogens and persistent pollutants. *Environ. Sci. Poll. Res.* **2020**, *27*, 9669–9685. [CrossRef]
77. Bugg, T.D.; Williamson, J.J.; Rashid, G.M. Bacterial enzymes for lignin depolymerisation: New biocatalysts for generation of renewable chemicals from biomass. *Curr. Opin. Chem. Biol.* **2020**, *55*, 26–33. [CrossRef]
78. Chauhan, P.S. Role of various bacterial enzymes in complete depolymerization of lignin: A review. *Biocatal. Agric. Biotech.* **2020**, *23*, 101498. [CrossRef]
79. Voß, H.; Heck, C.A.; Schallmey, M.; Schallmey, A. Database mining for novel bacterial β-etherases, glutathione-dependent lignin-degrading enzymes. *Appl. Environ. Microb.* **2020**, *86*, e02026-19. [CrossRef] [PubMed]
80. Bano, K.; Kuddus, M.; R Zaheer, M.; Zia, Q.; F Khan, M.; Gupta, A.; Aliev, G. Microbial enzymatic degradation of biodegradable plastics. *Curr. Pharm. Biotech.* **2017**, *18*, 429–440. [CrossRef]
81. Mohanan, N.; Montazer, Z.; Sharma, P.K.; Levin, D.B. Microbial and enzymatic degradation of synthetic plastics. *Front. Microbiol.* **2020**, *11*, 580709. [CrossRef]
82. Banik, J.; Chakraborty, D.; Rizwan, M.; Shaik, A.H.; Chandan, M.R. Review on disposal, recycling and management of waste polyurethane foams: A way ahead. *Waste Manag. Res.* **2023**, *41*, 1063–1080. [CrossRef]
83. Iftikhar, S.; Saleem, M.; Ahmad, K.S.; Jaffri, S.B. Synergistic mycoflora–natural farming mediated biofertilization and heavy metals decontamination of lithospheric compartment in a sustainable mode via *Helianthus annuus*. *J. Int. Environ. Sci. Tech.* **2019**, *16*, 6735–6752. [CrossRef]
84. Iram, S.; Basri, R.; Ahmad, K.S.; Jaffri, S.B. Mycological assisted phytoremediation enhancement of bioenergy crops *Zea mays* and *Helianthus annuus* in heavy metal contaminated lithospheric zone. *Soil Sediment Contam. Int. J.* **2019**, *28*, 411–430. [CrossRef]
85. Iram, S.; Iqbal, A.; Ahmad, K.S.; Jaffri, S.B. Congruously designed eco-curative integrated farming model designing and employment for sustainable encompassments. *Environ. Sci. Poll. Res.* **2020**, *27*, 19543–19560. [CrossRef]
86. Montazer, Z.; Habibi Najafi, M.B.; Levin, D.B. Challenges with verifying microbial degradation of polyethylene. *Polymers* **2020**, *12*, 123. [CrossRef]
87. Yuan, J.; Ma, J.; Sun, Y.; Zhou, T.; Zhao, Y.; Yu, F. Microbial degradation and other environmental aspects of microplastics/plastics. *Sci. Total Environ.* **2020**, *715*, 136968. [CrossRef]
88. Dobranic, J.K.; Zak, J.C. A microtiter plate procedure for evaluating fungal functional diversity. *Mycologia* **1999**, *91*, 756–765. [CrossRef]
89. Mathur, G.; Prasad, R. Degradation of polyurethane by *Aspergillus flavus* (ITCC 6051) isolated from soil. *Appl. Biochem. Biotech.* **2012**, *167*, 1595–1602. [CrossRef] [PubMed]
90. Holt, J.G.; Krieg, N.; Sneath, P.H.A.; Staley, J.T.; Williams, S.T. *Bergey's Manual of Determinative Bacteriology*, 9th ed.; Williams & Wilkins: Baltimore, PA, USA, 1994; ISBN 0-683-00603-7.
91. Shah, A.A.; Hasan, F.; Akhter, J.I.; Hameed, A.; Ahmed, S. Degradation of polyurethane by novel bacterial consortium isolated from soil. *Annals Microb.* **2008**, *58*, 381–386. [CrossRef]
92. Liu, J.; He, J.; Xue, R.; Xu, B.; Qian, X.; Xin, F.; Blank, L.M.; Zhou, J.; Wei, R.; Dong, W.; et al. Biodegradation and up-cycling of polyurethanes: Progress, challenges, and prospects. *Biotechnol. Adv.* **2021**, *48*, 107730. [CrossRef]
93. Raaman, N.; Rajitha, N.; Jayshree, A.; Jegadeesh, R. Biodegradation of plastic by *Aspergillus* spp. isolated from polythene polluted sites around Chennai. *J. Acad. Indus. Res.* **2012**, *1*, 313–316.
94. Ibrahim, I.N.; Anwar, M.; Khalid, M.H.; Ismail, M.S.; Hamzah, M.M.; Toshiaki, N.K. Polyester-polyurethane biodegradation by *Alternaria Solani*, isolated from Northern Jordan. *Adv. Environ. Biol.* **2009**, *3*, 162–170.
95. Gajendiran, A.; Krishnamoorthy, S.; Abraham, J. Microbial degradation of low-density polyethylene (LDPE) by *Aspergillus clavatus* strain JASK1 isolated from landfill soil. *3 Biotech* **2016**, *6*, 52. [CrossRef]
96. Wang, Y.; Luo, L.; Li, X.; Wang, J.; Wang, H.; Chen, C.; Guo, H.; Han, T.; Zhou, A.; Zhao, X. Different plastics ingestion preferences and efficiencies of superworm (*Zophobas atratus Fab.*) and yellow mealworm (*Tenebrio molitor* Linn.) associated with distinct gut microbiome changes. *Sci. Total Environ.* **2022**, *837*, 155719. [CrossRef]
97. Luo, L.; Wang, Y.; Guo, H.; Yang, Y.; Qi, N.; Zhao, X.; Gao, S.; Zhou, A. Biodegradation of foam plastics by *Zophobas atratus* larvae (Coleoptera: Tenebrionidae) associated with changes of gut digestive enzymes activities and microbiome. *Chemosphere* **2021**, *282*, 131006. [CrossRef]
98. Porter, R.; Černoša, A.; Fernández-Sanmartín, P.; Martínez Cortizas, A.; Aranda, E.; Luo, Y.; Zalar, P.; Podlogar, M.; Gunde-Cimerman, N.; Gostinčar, G. Degradation of polypropylene by fungi *Coniochaeta hoffmannii* and *Pleurostoma richardsiae*. *Microbiol. Res.* **2023**, *277*, 127507. [CrossRef]

99. Zhu, P.; Gong, S.; Deng, M.; Xia, B.; Yang, Y.; Tang, J.; Qian, G.; Yu, F.; Goonetilleke, A.; Li, X. Biodegradation of waste refrigerator polyurethane by mealworms. *Front. Environ. Sci. Eng.* **2023**, *17*, 38. [CrossRef]
100. Salgado, C.A.; Pereira Vidigal, P.A.; Dantas Vanetti, M.C. Biodegradation of polyurethanes by *Staphylococcus warneri* and by microbial co-culture. *Chemosphere* **2024**, *359*, 142169. [CrossRef] [PubMed]
101. Weng, Y.; Han, X.; Sun, H.; Wang, J.; Wang, Y.; Zhao, X. Effects of polymerization types on plastics ingestion and biodegradation by *Zophobas atratus* larvae, and successions of both gut bacterial and fungal microbiomes. *Environ. Res.* **2024**, *251*, 118677. [CrossRef] [PubMed]
102. Luo, G.; Hou, Z.; Tian, L.; Tan, H. Comparison of nitrate-removal efficiency and bacterial properties using PCL and PHBV polymers as a carbon source to treat aquaculture water. *Aquacul. Fish.* **2020**, *5*, 92–98. [CrossRef]
103. Cárdenas Espinosa, M.J.; Blanco, A.C.; Schmidgall, T.; Atanasoff-Kardjalieff, A.K.; Kappelmeyer, U.; Tischler, D.; Eberlein, C. Toward biorecycling: Isolation of a soil bacterium that grows on a polyurethane oligomer and monomer. *Front. Microbiol.* **2020**, *11*, 404. [CrossRef]
104. Gu, J.-D. Microbiological deterioration and degradation of synthetic polymeric materials: Recent research advances. *Int. Biodeter. Biodegrad.* **2003**, *52*, 69–91. [CrossRef]
105. Gu, J.-D. Biodegradability of plastics: The issues, recent advances, and future perspectives. *Environ. Sci. Pollut. Res.* **2021**, *28*, 1278–1282. [CrossRef]
106. Gu, J.-G.; Gu, J.-D. Methods currently used in testing microbiological degradation and deterioration of a wide range of polymeric materials with various degree of degradability: A review. *J. Polym. Environ.* **2005**, *13*, 65–74. [CrossRef]
107. Kyaw, B.M.; Champakalakshmi, R.; Sakharkar, M.K.; Lim, C.S.; Sakharkar, K.R. Biodegradation of low density polythene (LDPE) by *Pseudomonas* species. *J. Ind. Microbiol.* **2012**, *52*, 411–419. [CrossRef]
108. Cárdenas Espinosa, M.J.; Schmidgall, T.; Wagner, G.; Kappelmeyer, U.; Schreiber, S.; Heipieper, H.J.; Eberlein, C. An optimized method for RNA extraction from the polyurethane oligomer degrading strain Pseudomonas capeferrum TDA1 growing on aromatic substrates such as phenol and 2,4-diaminotoluene. *PLoS ONE* **2021**, *16*, e0260002. [CrossRef]
109. Kopteva, Z.O.; Vortman, M.Y.; Iutynska, G.O.; Kopteva, G.E.; Abdulina, D.R.; Lemeshko, V.M.; Terebilenko, A.V.; Pylypenko, A.M.; Shevchenko, V.V. Microbial destruction of guanidine-containing polymers. *Mater. Sci.* **2023**, *59*, 494–503. [CrossRef]
110. Loredo-Treviño, A.; Gutiérrez-Sánchez, G.; Rodríguez-Herrera, R.; Aguilar, C.N. Microbial enzymes involved in polyurethane biodegradation: A review. *J. Polymers Environ.* **2012**, *20*, 258–265. [CrossRef]
111. Munir, E.; Harefa, R.S.M.; Priyani, N.; Suryanto, D. Plastic degrading fungi *Trichoderma viride* and *Aspergillus nomius* found in Indonesia. *J. Phys. Conf. Ser.* **2018**, *953*, 012025. [CrossRef]

polymers

MDPI

Article

Analyzing Thermal Degradation Effects on Devulcanized GTR-Based NR/SBR/NBR Rubber Compounds Reinforced with SiO_2 Particles

Xavier Colom [1,*], Laia Farrés [1], Ramon Mujal [2], Shifeng Wang [3] and Javier Cañavate [1]

1 Department of Chemical Engineering, Universitat Politècnica de Catalunya Barcelona Tech, C/Colom, 1, 08222 Terrassa, Barcelona, Spain; laia.farres@estudiantat.upc.edu (L.F.); francisco.javier.canavate@upc.edu (J.C.)

2 Department of Electrical Engineering, Universitat Politècnica de Catalunya Barcelona Tech, C/Colom, 1, 08222 Terrassa, Barcelona, Spain; mujal@ee.upc.edu

3 School of Chemistry and Chemical Engineering, Shanghai Jiao Tong University, Shanghai 200240, China; shfwang@sjtu.edu.cn

* Correspondence: xavier.colom@upc.edu

Abstract: The large number of tires produced annually demands new recycling methods. A key challenge associated with recycling elastomers is their crosslinking structure that prevents fusion. It is possible to reverse crosslinking through a process called devulcanization. Devulcanized elastomers can be blended with fresh rubber and revulcanized for reuse. This paper examines samples made from natural rubber (NR), styrene–butadiene rubber (SBR), and nitrile butadiene rubber (NBR), blended with varying proportions of devulcanized ground tire rubber (dGTR) and newly revulcanized rubber. SiO_2, commonly present in tire formulations, is also added. Samples of these materials, with 0, 10, 20, and 40 phr of dGTR are subjected to accelerated degradation for 0, 30, 60, 120, and 240 h. The effects of this treatment, the influence of SiO_2, and the presence of a silane-based devulcanization agent (TESPT) that promotes the interaction between the rubber and silica, are analyzed at the microstructural level (FTIR, TGA, SEM) and through mechanical properties testing. The microstructural results of the spectroscopy and thermal analysis show that interactions between dGTR, silica, and silane compounds form aggregates that impact the material properties and degradation of the tires. Mechanically, when the sample contained up to 20 phr of dGTR, the compound presented a more brittle behavior, due to the crosslinking induced by the degradation.

Keywords: ground tire rubber (GTR); natural rubber (NR); styrene–butadiene rubber (SBR); nitrile butadiene rubber (NBR); SiO_2; TESPT; devulcanization process

1. Introduction

Green rubber compounds present a promising solution to recover a significant portion of the one billion end-of-life tires generated globally each year [1]. In Europe alone, around 4.2 million tons of used tires were generated in 2023, according to the European Tyre and Rubber Manufacturers' Association (ETRMA) [2]. However, only 600,000 tons are reused or retreaded, leaving 3.6 million tons classified as end-of-life tires. This substantial accumulation of used tires highlights the urgent need for valorization strategies to: (i) reduce landfill waste, (ii) prevent fires, (iii) conserve natural resources, (iv) lower water and energy consumption, (v) minimize the carbon footprint, and (vi) promote a sustainable economy, while protecting biodiversity.

As residua, used tires are especially difficult to deal with because they are bulky, non-biodegradable, and occupy significant space in landfill sites, where they generate leachates and release harmful chemicals, posing serious environmental risks. Additionally, they can become breeding grounds for disease-carrying mosquitoes and are highly

flammable, increasing the risk of fires that emit toxic gases and present a significant threat to public health.

Furthermore, tire recycling should be a more energy-efficient process than manufacturing new tires, leading to significant reductions in fossil fuel consumption and greenhouse gas emissions. Producing one ton of new tires requires 21.7 MWh of energy, whereas recycling this amount of tires would likely offer substantial energy savings in comparison to this data, helping to reduce the industry's carbon footprint. According to the United States Tire Manufacturers Association (USTMA) [3], in the event of increased tire recycling, CO_2 emissions would drop from 63.64 kg to 30.58 kg per ton of tires, representing a 52% reduction, while the demand for virgin raw materials would decrease by 45%, conserving resources and reducing the reliance on non-renewable sources.

Among tire recycling methods, devulcanization has gained particular attention because of its conception based on breaking the crosslinked structure of elastomers, by cleaving the sulfur bonds formed during vulcanization [4,5]. Devulcanization enables the reprocessing of rubber, allowing it to be blended with new materials, such as styrene–butadiene rubber (SBR), natural rubber (NR), and nitril rubber (NBR), which is common in tire production. The blend can be posteriorly revulcanized again. This process offers a way to reintegrate waste rubber into new tire manufacturing, while maintaining similar material properties. Ground tire rubber (GTR), a powder produced from recycled car and truck tires, is generally the starting material in the devulcanization process [6]. The subsequent product, devulcanized GTR (dGTR) is then blended and revulcanized with the other components in order to produce tire rubber [7].

The properties of blends incorporating dGTR and the degradation that tires undergo during their service life are still underexplored. This degradation can cause significant changes in the rubber's properties and these alterations may affect the performance of new tires made with devulcanized GTR. Understanding the impact of this degradation is essential to ensure the viability of using devulcanized rubber in tire manufacturing [8].

The efforts to reduce CO_2 emissions and advance towards a more sustainable economy are leading to the adoption of electric vehicles (EVs). EVs produce no direct emissions, which lowers their carbon footprint. However, due to the increased weight of electric cars, tire formulations have had to be adapted. A significant proportion of silica (up to 30 phr) has been incorporated into tires for electric vehicles. Studies show that the addition of silica (SiO_2), high-defect multiwalled carbon nanotubes (HD-CNT), and carbon black (CB), in the formulation of these elastomeric materials, enhances their mechanical properties, thermal conductivity, and volume resistivity [9]. Silica, in particular, offers benefits, such as improved fuel efficiency and better grip, especially in wet conditions [10,11].

The challenge with using an increased proportion of silica nanoparticles is their poor dispersion within the elastomeric matrix, primarily due to the differences in polarity between the two materials that can result in the formation of agglomerates, reducing the effectiveness of silica reinforcement. To address this issue, silanes are commonly used as coupling agents. These compounds have functional groups that bond with both the elastomer matrix and the silica particles, improving their dispersibility and enhancing the overall performance [12–14].

In this study, the POLQUITEX research group, with proven experience in polymer aging processes [15], aims to investigate vulcanized blends that mimic the composition of a tire that incorporates 0, 10, 20, and 40 phr of devulcanized ground tire rubber (dGTR), styrene–butadiene rubber (SBR), natural rubber (NR), and nitril rubber (NBR). These blends are also degraded for 0, 30, 60, 120, and 240 h. The effect of different dGTR content and compound compositions, the impact of the addition of SiO_2, the changes in the tensile behavior, and the influence of degradation over time, will be analyzed.

2. Materials and Methods

2.1. Materials

NR SVR CV (Vietnam natural rubber), with ash content at a max. of 0.4% wt, volatile matter at 0.8% wt, with a Plasticity Retention Index (PRI) of 60, and a density of 0.92 g/cm^3; SBR Pliogum 1027K1, with 24 wt.% of bonded styrene (MV ML1+4, 100 °C; 52); and NBR Ker N-29, with 29 wt.% of bonded acrylonitrile (MV ML1+4, 100 °C; 50); were supplied by VIGAR HEXPOL Compounding SLU, Rubí (Spain).

Ground tire rubber (GTR) is obtained from the shredding of the tires of vehicles in special processing plants. In this case, the GTR is a mix of passenger and truck tires, with particle sizes of up to 0.7 mm and an average particle size of 320 microns, which were received by Gestión Medioambiental de Neumáticos S. L. (GMN) in Maials (Spain). The composition of the rubber, according to our thermogravimetric analysis, is natural rubber 29%; synthetic rubber 31%; carbon black 29.5%; SiO_2 7%; other additives (e.g., curing system, processing aids) 3.5%; which is very close to the composition of the rubber obtained by Li et al. [16]. Details on the particle size distribution have been published previously by Colom et al. [17].

The additives and curing agents used for curing, namely vulcanization accelerators (TBBS N-cyclohexil-2-benzotiazolsulfenamide, TMTD—tetramethylthiuram disulfide), carbon black N550, stearic acid, zinc oxide, sulfur with technical purity, and SiO_2 (85% particle size below 150 microns), were supplied by VIGAR Rubí (Spain).

Bis (3-triethoxysilyl propyl) tetrasulfide (TESPT) activated the devulcanization. It was also supplied by VIGAR Rubí (Spain).

2.2. Devulcanization of GTR

The devulcanization process for GTR using combined thermo-mechanical/chemical procedures was carried out by modifying a procedure already tested in similar cases and published by our research group in a previous paper [18].

The thermo-mechanical/chemical treatment is carried out using GTR, with the addition of 2 parts per hundred rubber (phr) of TESPT, which acts as a devulcanizing agent [19]. This blend is passed through the machine 10 times at 50 °C and 80 rpm. The sample, after this treatment, presents as putty-like when touched, an aspect which is often observed in devulcanized rubber.

2.3. Preparation of Rubber Compounds

The rubber compounds were prepared using the formulation presented in Table 1. The first step in the mixing process consists of the preliminary plasticization of the fresh elastomeric compounds, NR, SBR, and NBR, for 5 min. After that, the mixing stage takes place, involving devulcanized GTR, 30 phr of carbon black, and 30 phr of SiO_2, for 5 min. Finally, the sulfur curing system (zinc oxide 5.0 phr; stearic acid 3.0 phr; TBBS 1.0 phr; TMTD 0.25 phr; sulfur 2.0 phr) is added to the blend and mixed for 2 min, allowing the revulcanization process to take place.

Table 1. Sample codes and composition of different compounds in phr (per hundred rubber).

Sample Code	NR	SBR	NBR	dGTR	SiO_2	CB	S	ZnO	HSt	TBBS	TMTD
0GTRNRSBRNBR	50	35.0	15.0	0	30.0	30	2	5	3	1	0.25
10GTRNRSBRNBR	47	33.0	14.0	10	29.3	27	2	5	3	1	0.25
20GTRNRSBRNBR	44	31.0	13.0	20	28.6	24	2	5	3	1	0.25
40GTRNRSBRNBR	38	27.0	11.0	40	27.2	20	2	5	3	1	0.25

The rationale behind the studied formulations is to maintain the overall composition of a typical tire, while substituting only the elastomers with those found in devulcanized

ground tire rubber (dGTR). The baseline tire composition is defined in the initial formulation. For example, when 10 phr of dGTR is added, it introduces not only elastomers, in the same initial proportion, but also, according to the composition of a typical tire, 0.7 phr of SiO_2 and 3 phr of carbon black (CB). To preserve the original proportions, we adjust these components accordingly, reducing SiO_2 by 0.7 phr and CB by 3 phr for every 10 phr of dGTR added. The remaining components, associated with the curing system, are unaffected by the addition of dGTR and are incorporated in the same proportions as in the original blend.

The rubber ingredients were mixed in a Plasticorder internal mixer (Brabender GmbH & Co. KG, Duisburg, Germany) at 80 °C and 80 rpm. The order of appearance of the components presented in Table 1 (left to right) also reflects the order of mixing. The devulcanized GTR content in the samples was 0, 10, 20, and 40 parts per hundred rubber (phr).

Two batches of the mix, cited above, are necessary to create a suitable sample plate of the polymer. These two batches are recombined in a two-roll mill Collin Teach-line press. The resulting material is subsequently molded in a Collin P 200E hot plate press at 160 °C and 200 bar for 12 min [20]. These conditions correspond to the optimum curing time (t90), determined by the vulcanization characteristics of vulcanizable rubber composites, according to ASTM D2084 [21], with a Monsanto Oscillating Disc Rheometer R 100 (Akron, OH, USA), at 162 ± 1 °C. After this time, the mold is removed carefully from the press, using Kevlar gloves, and quickly taken away to cool. Finally, the mold is separated from the part, which is now ready to shape into the test samples with a J.BOT Instruments S.A. (Barcelona, Spain) test cutter, designed to obtain dumbbell-shaped test samples, according to the specifications in ASTM D-412 [22].

2.4. Thermo-Oxidation Process

The aging treatment was performed in an air-forced oven at 80 °C (Selecta, Barcelona, Spain). The samples were exposed to the conditions for different time periods (0, 30, 60, 120, and 240 h), according to ISO 4577 [23].

2.5. Measurements

The tensile properties of the vulcanized rubber dGTR/NR/SBR/NBR composites were evaluated according to the ISO 37 [24]. Testing was conducted using a high-precision Instron 3366 testing machine from Norwood, MA (USA), with a pneumatic cell load capacity of 2 kN. The tensile tests were executed at a constant cross-head speed of 500 mm/min, in a controlled environment, with a relative humidity (RH) of 50 ± 5% and a temperature of 23 ± 2 °C. Five specimens per test were used.

The chemical structure of the samples was determined using FTIR–ATR analysis, performed by means of a Spectrum Two spectrometer from Perkin Elmer (Waltham, MA, USA). The device had a diamond ATR crystal attachment. The spectra were registered at a resolution of 2 cm^{-1} and 40 scans took place in the range of 3500–500 cm^{-1}, according to which the compound signals related to different deformation bands can be observed.

The TGA was performed using Perkin Elmer TGA 8000 apparatus (USA). A sample of the composites, weighing approximately 10 mg, was placed in a corundum dish. The measurement was conducted in the temperature range 30–800 °C and under atmospheric oxidation with air (30 mL/min), at a heating rate of 20 °C/min.

The fracture surface of the samples, as a result of mechanical testing, was examined with a JEOL JSM 6480 scanning electron microscope from Akishima, Tokyo, Japan. The dry samples were coated with a thin layer of gold before observation under the microscope, to increase the sample conductivity.

3. Results

3.1. Mechanical Properties

As can be seen in Figure 1, Young's modulus (YM) follows an increasing trend as a function of the exposure time to thermo-oxidative aging. Likewise, it can also be observed

that for the same exposure time, the trends change as a function of the dGTR content of the samples. Some of the samples with an exposure time of 120 h, have YM values that show increases of 140%. This is due to two phenomena, namely the formation of crosslinks in the thermo-oxidated elastomeric chains and the generation of hard and brittle layers around the particles, as has been reported in previous studies [25–27].

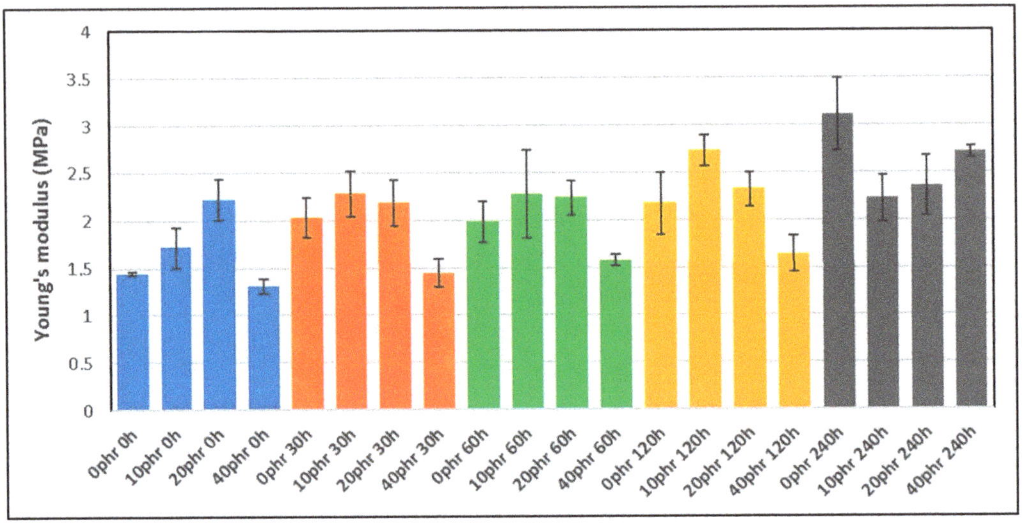

Figure 1. Young's modulus (YM) of different compounds as a function of dGTR content and thermo-oxidative exposure time.

The brittleness of these samples is due to different processes: (a) the devulcanization of the GTR particles in the presence of silane-based TESPT induces a significant affinity of the devulcanized GTR to the SiO_2 micro-fillers, forming an aggregate of TESPT–SiO_2 around the dGTR particles; (b) this aggregate prevents, at least partially, the exudation of carbon black particles to the surface, caused by the increase in temperature to 80 °C, and they are placed between the interstitial spaces of the TESPT–SiO_2, forming a hard and brittle layer; and (c) when the dGTR content is lower than 20 phr, YM grows significantly, above this the samples present brittle behavior that decreases the mechanical properties. With the increased exposure time, the amount of CB diffused increases and, consequently, the thickness of the layer around the dGTR also increases.

Figure 2 presents the tensile strength (TS) values as a function of the thermo-oxidative aging exposure time and dGTR content (phr). The results show an increase in the TS up to 60 h of exposure, with dGTR content of 20 phr. It is interesting to see how, in all cases, a dGTR content of 40 phr causes a significant decrease in the TS. This is due to the presence of large agglomerates of dGTR wrapped in TESPT–SiO_2 and CB, which generate a stress nucleus, where the samples easily break, and are produced when the content of dGTR increases. It can also be concluded that aging for 60 h does not alter the TS. Comparing these samples (dGTR/NR/SBR/NBR/SiO_2) with the samples dGTR/NBR/SiO_2, it can be seen that they have similar tensile properties [26]. It should also be considered that when the size of the dGTR particles is large, the affinity between the elastomeric matrix and the dGTR is low and causes premature breakage of the samples [27].

The elongation at break (EB) as a function of the aging time (Figure 3) follows a fairly defined decreasing trend. Higher EB values are observed for the samples with 0 and 10 phr dGTR, up to an exposure time of 60 h. Between 120 and 240 h, the decrease in the elongation at break is significant. In samples with 20 and 40 phr dGTR, a marked decrease is also

observed, mainly for the non-aged samples. The behavior at the EB is also explained by the same phenomena discussed previously.

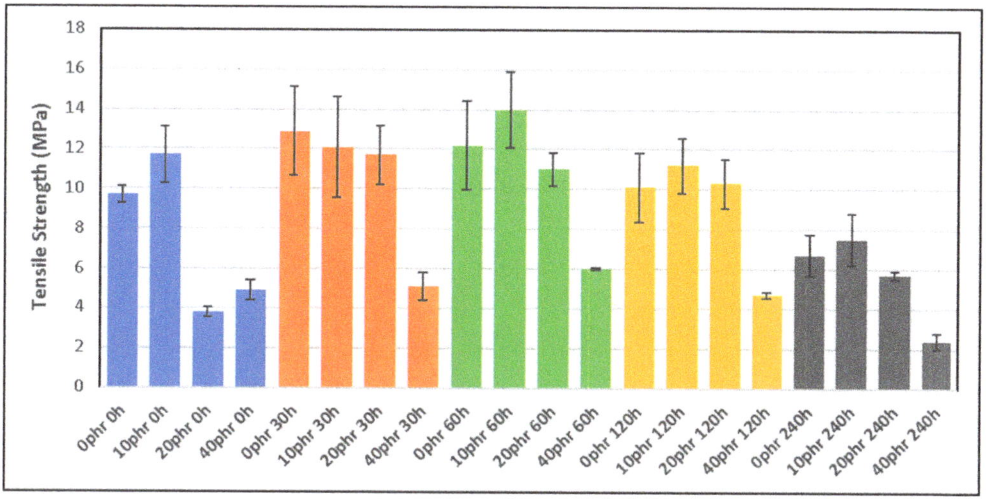

Figure 2. Tensile strength (TS) of different compounds as a function of dGTR content and thermo-oxidative exposure time.

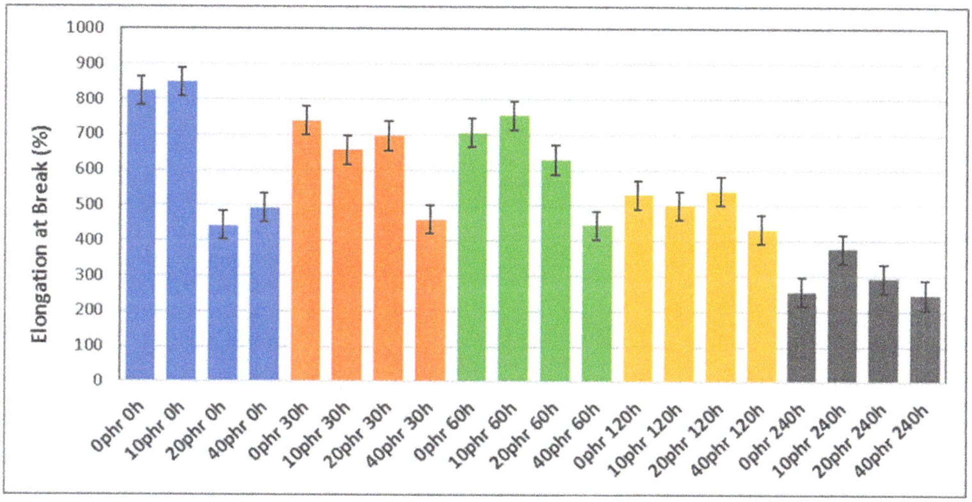

Figure 3. Elongation at break (EB) of different compounds as a function of dGTR content and thermo-oxidative exposure time.

The diagram in Figure 4 shows the different phases of the phenomenon that takes place in the compounds without and the compounds containing dGTR particles, neat elastomers (NR, SBR, and NBR), 30 phr of CB, and 30 phr of SiO_2. It can be observed how the dGTR, due to the presence of TESPT, retains SiO_2 micro-fillers in a wrapped manner, which at the same time prevents the exudation of CB towards the surface, forming SiO_2–CB agglomerates, which link multiple particles around the dGTR. These larger aggregates significantly weaken the structure of the sample [28].

(a) (b)

Figure 4. Proposal squema showing the phenomenon of CB and SiO_2 migration from the core to the surface in: (**a**) the NR/SBR/NBR composite without dGTR and (**b**) composites with dGTR. Blue strands correspond to SBR, yellow strands to NR, lilac strands to NBR, black dots to CB, and blue dots to SIO_2. The enlarged red circle shows the SiO_2 evolved dGTR particle that prevents CB exudation, generating a brittle layer of CB.

3.2. FTIR Spectroscopy Characterization

As can be observed in the spectra in Figure 5, the most interesting bands to be analyzed are those corresponding to SiO_2 at 1057 cm^{-1} and carbon black (CB) at 1016 cm^{-1}. The samples without dGTR present an intense band corresponding to SiO_2 in the unaged samples. The band assigned to CB at 1016 cm^{-1} is intense and is not altered by the degradation time. As mentioned in regard to the measurements, the spectra were recorded by ATR spectroscopy and, as such, are representative of the surface of the samples. The difference in the variations in the intensity of those two bands is then related to the diffusion phenomena. The band assigned to CB remains at a high and constant intensity because the nanometric size of this filler gives it a high capacity to diffuse through elastomeric structures in order to reach the surface; the amount of CB observed in the spectra is not altered by the degradation time. Instead, the SiO_2 crystals, originally dispersed within the elastomeric matrix, diffuse towards the surface with a longer degradation time. After 60 h, the elastomeric structures of NR (which are less stable than those of SBR and NBR) [29] begin to degrade, generating a greater number of branches, oxidation groups, and crosslinks. These structural changes, together with the abovementioned effect of TESPT attaching, prevent greater diffusion of SiO_2 towards the surface, which is observed with a lower band intensity.

The band at 1538 cm^{-1}, assigned to $ZnSt_2$, increases its intensity with long aging times. The diffusion of $ZnSt_2$ into similar elastomeric samples has been studied previously [30,31]. The particles of $ZnSt_2$, at a degradation temperature of 80 °C, can diffuse towards the surface and the degradation of the material improves the particles' ability to move through the structures. The other bands, 960, 790, 697, and 451 cm^{-1}, are assigned to -CH=CH-, trans-polybutadiene, benzenic rings, and S-S, respectively. The intensity of each band depends on the number of groups that appear at the surface [32–35].

There is an interesting effect in regard to the SiO_2 band at 1057 cm^{-1} that further corroborates the previous discussion and is present in Figure 6. When dGTR is added to the rubber compound (RC) formulation, the band at 1057 cm^{-1} shifts to 1076 cm^{-1}, and its intensity decreases for all aging periods (Figure 6). This is a consequence of the adhesion or wrapping effect on the SiO_2 particles caused by the affinity produced by the coupling

agent TESPT, which causes the adhesion of the SiO_2 particles to the surface of the dGTR. Furthermore, according to Ghorai et al. [19], devulcanized GTR naturally has much more of an affinity for SiO_2 particles than for other particles in the sample, such as CB or $ZnSt_2$. Likewise, as represented in the squama in Figure 4, both particles can be retained by the SiO_2 aggregates that form around the dGTR. The spectral intensity of these bands (CB and $ZnSt_2$) decrease in the RC with dGTR, in comparison with the RC without dGTR.

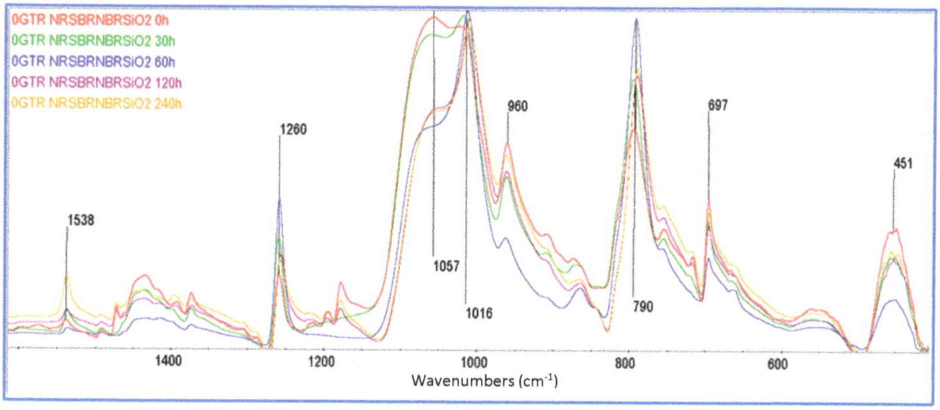

Figure 5. FTIR spectra of NR/SBR/NBR/SiO_2 compounds as a function of different thermo-oxidation exposure times (0, 30, 60, 120, and 240 h). Y axis in arbitrary units.

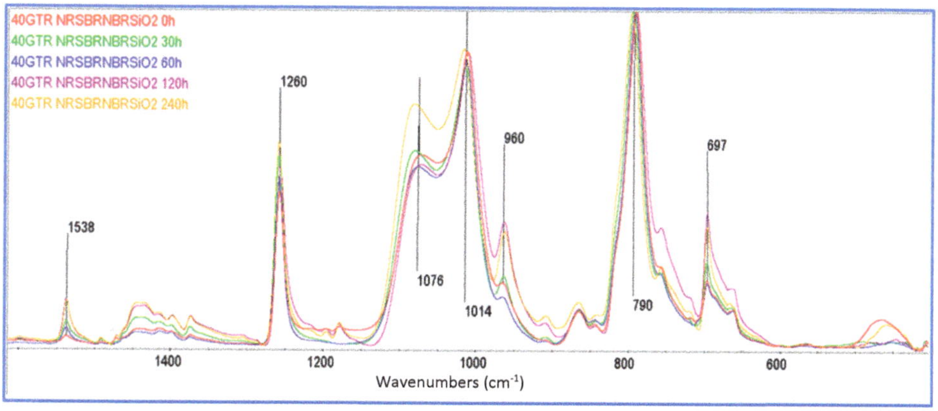

Figure 6. FTIR spectra of dGTR/NR/SBR/NBR/SiO_2 compounds filled with 40 phr of dGTR as a function of different thermo-oxidation exposure times (0, 30, 60, 120, and 240 h). Y axis in arbitrary units.

3.3. Thermogravimetric Analysis

Figure 7a shows the TGA thermograms of the dGTR/NR/SBR/NBR/SiO_2 compounds, illustrating various dGTR and silica loadings. The weight loss percentages of the samples with different dGTR content are shown, revealing a consistent residue value of 20%, attributed to SiO_2. The DTGA results indicate a heterogeneous elastomeric composition (NR/SBR/NBR), where each component exhibits distinct thermal decomposition peaks at different temperatures. In samples with low dGTR content (0% and 10%), three peaks are identifiable: the first corresponds to NR (370 °C), the second to SBR (445 °C), and the third to NBR (455 °C). As the dGTR content increases, the thermal decomposition

temperature (TdT) of SBR and NBR join into a single domain above 450 °C, while NR remains clearly above 400 °C. It is also worth noting that the TdT of different components increases with silica [36].

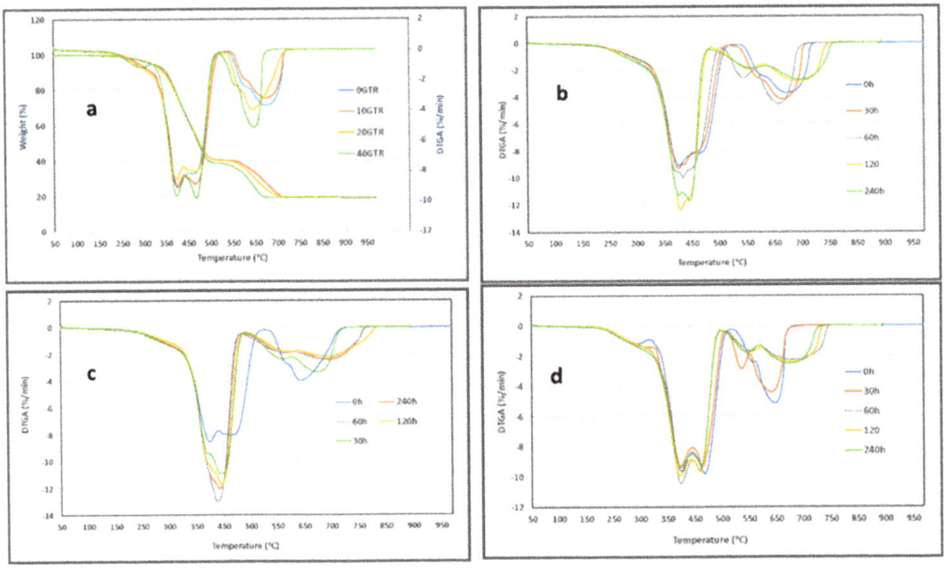

Figure 7. (**a**) TGA and DTGA of dGTR/NR/SBR/NBR/SiO_2 compounds filled with 0, 10, 20, and 40 phr of dGTR; (**b–d**) DTGA of dGTR/NR/SBR/NBR/SiO_2 compounds filled with 0, 20, and 40 phr of dGTR as a function of different thermo-oxidation exposure times (0, 30, 60, 120, and 240 h).

NR and NBR are incompatible, resulting in the presence of distinct domains for both polymers in the compound. However, there is partial miscibility between NR and NBR (due to presence of SBR), leading to interactions and potential crosslinking of the polymer chains, which causes a shift in the peak temperatures. This is particularly evident for the NBR peak, which shifts from 485 °C (without NR) to approximately 460 °C, for a composition of 67% NR and 33% NBR [8].

As the dGTR amount increases, the significance of these peaks also rises. This is attributed to the dGTR, derived from recycled tires, which contains virgin elastomers that enhance the overall amount of elastomers in the sample. However, the degradation temperatures across the samples do not show substantial differences. The subsequent peak corresponds to carbon black (CB) [12], whose decomposition temperature varies among the different samples. Remarkably, samples with higher dGTR content (higher than 20% dGTR) decompose first, unlike those with lower dGTR amounts (0% and 10%). This trend is likely due to the increased formation of aggregates in the samples with higher dGTR, which reduces the degree of dispersion and significantly reduces the TdT range. The aggregates further associate into even larger structures, via van der Waals forces, to form agglomerates [11].

Figure 7b–d illustrates the DTGA curves of the 0, 20, and 40 phr dGTR thermo-oxidated compounds, revealing three distinct peaks. As previously mentioned, the first peak corresponds to NR, the second to the SBR–NBR domain, and the third to the TdT of carbon black (CB). The thermo-oxidation process significantly modifies these peaks, with the samples containing lower dGTR content (0 and 20 phr) being more affected than those with 40 phr (Figure 7d). This is because thermo-oxidation at 80 °C is more effective for samples with a higher raw elastomer content (NR, SBR, and NBR), compared to those with greater dGTR content.

Particularly, there are substantial differences in the broad peak of the TdT for CB among the samples without dGTR (Figure 7b), those with 20 phr dGTR (Figure 7c), and those with 40 phr dGTR (Figure 7d). In the samples without dGTR, the dispersion of CB nanoparticles is more uniform, and their mobility and infiltration within the silica matrix increase with the temperature and prolonged exposure time (up to 120 days). This enhanced dispersion contributes to a wider range of TdTs for these particles.

Alternatively, in samples with dGTR, the agglomerates formed around the dGTR limit the effectiveness of thermo-oxidation at shorter times (30–60 h), resulting in a narrower TdT range. The presence of dGTR (using TESP as a devulcanizing agent) facilitates the formation of SiO_2 macroaggregates that encase the dGTR. In this scenario, more CB nanoparticles become entrapped within these macroaggregates, leading to a higher concentration of nanoparticles, which reduces the TdT range.

3.4. Scanning Electron Microscopy Analysis

The images in Figure 8 show the exudation process involving CB and SiO_2 that has been explained previously. These images correspond to the fractured surface of the samples. In Figure 8a, which corresponds to the compounds without dGTR, it is observed that the SiO_2 particles (more visible) have easily migrated to the surface of the sample. Figure 8b shows these particles in a maximized form and they correspond to SiO_2 [37]. Figure 8c,d is related to the compounds with 40 phr dGTR submitted to 0 h and 120 h of thermo-oxidative degradation, in which it is observed that the SiO_2 and CB particles are retained around the dGTR, forming agglomerates. In Figure 8d, this phenomenon is observed more intensely.

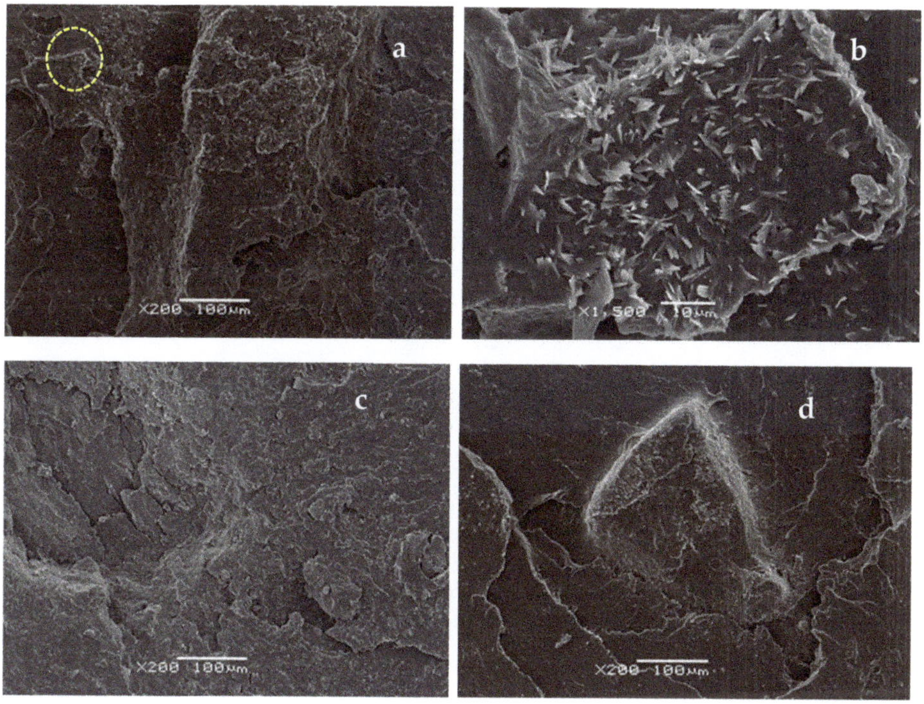

Figure 8. SEM photographs of fractured tensile compounds: (**a**) without dGTR, (**b**) magnification (×1500) of yellow circle, (**c**) with 40 phr dGTR at 0 h of thermo-oxidation, (**d**) with 40 phr dGTR at 120 h of exposure to the thermo-oxidation process.

4. Conclusions

The results of the present study show that the effect of degradation of dGTR-based, sustainable, NR/SBR/NBR rubber compounds affects the mechanical properties of the materials. Young's modulus increases with the thermo-aging exposure time, due to crosslink formation and the generation of hard, brittle layers around the dGTR particles. When the dGTR content exceeds 20 phr, the samples become more brittle. The tensile strength peaks at 60 h of exposure to thermo-aging in samples with 20 phr of dGTR content. Higher dGTR content (40 phr) leads to a decrease in the tensile strength. This is due to the effect of the silane-based devulcanizing agent that promotes interaction between the rubber and silica, generating large aggregates of dGTR wrapped in TESPT–SiO_2 and CB, which create a stress nucleus and weak points. The elongation at break decreases significantly with increasing dGTR content and prolonged aging. The samples with 0–10 phr of dGTR showed higher elongation, while those with 20–40 phr exhibited a marked decrease, particularly after 120 h of exposure to the thermo-oxidation process.

Using FTIR spectroscopy, the different SiO_2 and carbon black behaviors during degradation can be observed. SiO_2 diffusion decreases over time, while the diffusion of CB remains constant. SiO_2 particles form aggregates around the dGTR, due to the previously mentioned presence of a coupling agent (TESPT), which reduces diffusion, altering the mechanical properties of the materials. $ZnSt_2$ particles, as corroborated by other studies, show increased diffusion toward the surface with longer aging times.

The thermogravimetric analysis showed the patterns of decomposition of the involved elastomers and carbon black. The samples with higher dGTR content decomposed faster due to the formation of aggregates, which limited the dispersion and reduced the thermal stability of the materials. The samples with lower dGTR content exhibited distinct peaks for NR, SBR, and NBR, while the samples with higher dGTR content had merged thermal decomposition peaks, which indicates compatibility among the elastomers composing the samples. Aggregates formed by SiO_2-TESPT around the dGTR particles affect the dispersion of carbon black particles, which changes the thermal properties of the materials and results in a change to the thermal decomposition temperature.

Author Contributions: Conceptualization, X.C. and J.C.; methodology, X.C. and L.F.; validation, X.C., J.C., and S.W.; formal analysis, X.C., J.C., and L.F.; investigation, L.F. and R.M.; resources, X.C., J.C., and R.M.; data curation, X.C and J.C.; writing—original draft preparation, X.C.; writing—review and editing, X.C., J.C., and S.W.; visualization, X.C., J.C., R.M., and S.W; supervision, X.C.; project administration, X.C. and J.C.; funding acquisition, X.C. and J.C. All authors have read and agreed to the published version of the manuscript.

Funding: This research was funded as part of grant PID2021-126165OB-I00 funded by MCIN/AEI/10.13039/501100011033 and by ERDF, A way of making Europe, by the European Union.

Institutional Review Board Statement: Not applicable.

Data Availability Statement: The original contributions presented in the study are included in the article, further inquiries can be directed to the corresponding author.

Acknowledgments: We appreciate the collaboration and work carried out by Robert Blazquez and Patricia Muñoz from the Department of Chemical Engineering, Universitat Politècnica de Catalunya, BarcelonaTech, (ETSEIAAT) Terrassa, Spain.

Conflicts of Interest: The authors declare that there are no conflicts of interest.

References

1. Wiśniewska, P.; Haponiuk, J.T.; Colom, X.; Saeb, M.R. Green Approaches in Rubber Recycling Technologies: Present Status and Future Perspective. *ACS Sustain. Chem. Eng.* **2023**, *11*, 8706–8726. [CrossRef]
2. ETRMA. *Annual Report 2020/2021 and End Life Tyre Reports*; European Tire and Rubber Manufacturers Association: Bruxelles, Belgium, 2021.
3. USTMA (U.S. Tire Manufacturers Association). Available online: https://www.ustires.org/system/files/files/2024-09/USTMA_CTC_Manufacturing_sustainability.pdf (accessed on 18 July 2024).

4. Fazli, A.; Rodrigue, D. Recycling Waste Tires into Ground Tire Rubber (GTR)/Rubber Compounds: A Review. *J. Compos. Sci.* **2020**, *4*, 103. [CrossRef]
5. Movahed, S.O.; Ansarifar, A.; Estagy, S. Review of the reclaiming of rubber waste and recent work on the recycling of ethylene–propylene–diene rubber waste. *Rubber Chem. Technol.* **2016**, *89*, 54–78. [CrossRef]
6. Simon, D.Á.; Bárány, T. Effective Thermomechanical Devulcanization of Ground Tire Rubber with a Co-Rotating Twin-Screw Extruder. *Polym. Degrad. Stab.* **2021**, *190*, 109626. [CrossRef]
7. Molanorouzi, M.; Mohaved, S.O. Reclaiming waste tire rubber by an irradiation technique. *Polym. Degrad. Stab.* **2016**, *128*, 115–125. [CrossRef]
8. Zhao, J.; Yang, R.; Iervolino, R.; van der Vorst, B.; Barbera, S. The effect of thermo-oxidation on the continuous stress relaxation behavior of nitrile rubber. *Polym. Degrad. Stab.* **2015**, *115*, 32–37. [CrossRef]
9. Lu, Y.; Liu, J.; Hou, G.; Ma, J.; Wang, W.; Wei, F.; Zhang, L. From nano to giant? Designing carbon nanotubes for rubber reinforcement and their applications for high performance tires. *Compos. Sci. Technol.* **2016**, *137*, 94–101. [CrossRef]
10. Liu, Z. Rational Design of High-Performance Nanomaterials for Electric Vehicle Tires. *Highlights Sci. Eng. Technol.* **2023**, *62*, 110–118. [CrossRef]
11. Prasertsri, S.; Rattanasom, N. Fumed and precipitated silica reinforced natural rubber composites prepared from latex system: Mechanical and dynamic properties. *Polym. Test.* **2012**, *31*, 593–605. [CrossRef]
12. Young Ko, J.; Prakashan, K.; Kuk Kim, J. New silane coupling agents for silica tire treat compounds. *J. Elastomers Plast.* **2012**, *44*, 549–562.
13. Plueddemann, E.P. *Chemistry of Silane Coupling Agents*; Springer: Boston, MA, USA, 1991; pp. 31–54.
14. Ten Brinke, J.W.; Debnath, S.C.; Reuvekamp, L.A.E.M.; Noordermeer, J.W.M. Mechanistic aspects of the role of coupling agents in silica-rubber composites. *Compos. Sci. Technol.* **2003**, *63*, 1165–1174. [CrossRef]
15. Colom, X.; Cañavate, J.; Suñol, J.J.; Pagès, P.; Saurina, J.; Carrasco, F. Natural and artificial aging of polypropylene-polyethylene copolymers. *J. Appl. Polym. Sci.* **2002**, *87*, 1685–1692. [CrossRef]
16. Li, G.; Stubblefield, A.M.; Garrick, G.; Eggers, J.; Abadie, C.; Huang, B. Development of waste tire modified concrete. *Cem. Concret. Res.* **2004**, *34*, 2283–2289. [CrossRef]
17. Colom, X.; Marín-Genesca, M.; Mujal, R.; Formela, K.; Cañavate, J. Structural and physico-mechanical properties of natural rubber/GTR composites devulcanized by microwaves: Influence of GTR source and irradiation time. *J. Compos. Mater.* **2018**, *52*, 3099–3108. [CrossRef]
18. Colom, X.; Sans, J.; Bruijn, F.; Carrillo, F.; Cañavate, J. Structural, Thermal and Mechanical Assessment of Green Compounds with Natural Rubber. *Macromol* **2024**, *4*, 566–581. [CrossRef]
19. Ghorai, S.; Mondal, D.; Hait, S.; Kumar Ghosh, A.; Wiessner, S.; Das, A.; De, D. Devulcanization of Waste Rubber and Generation of Active Sites for Silica Reinforcement. *ACS Omega* **2019**, *4*, 18. [CrossRef]
20. Zedler, Ł.; Colom, X.; Cañavate, J.; Formela, K. GTR/NBR/Silica Composites Performance Properties as a Function of Curing System: Sulfur versus Peroxides. *Materials* **2021**, *14*, 5345. [CrossRef]
21. *ASTM D2084*; Standard Test Method For Rubber Property—Vulcanization Using Oscillating Disk Cure Meter. ASTM International: West Conshohocken, PA, USA, 2019.
22. *ASTM D412*; Standard Test Methods for Vulcanized Rubber and Thermoplastic Elastomers—Tension. ASTM International: West Conshohocken, PA, USA, 2008.
23. *ISO 4577*; Plastics—Polypropylene and Propylene-Copolymers—Determination of Thermal Oxidative Stability in Air—Oven Metho. International Organization for Standardization: Vernier, Geneva, 2012.
24. *ISO 37*; Rubber, Vulcanized or Thermoplastic—Determination of Tensile Stress-Strain Properties. International Organization for Standardization: Vernier, Geneva, 2017.
25. Song, P.; Wan, C.; Xie, Y.; Formela, K.; Wang, S. Vegetable derived-oil facilitating carbon black migration from waste tire rubbers and its reinforcement effect. *Waste Manag.* **2018**, *78*, 238–248. [CrossRef]
26. Cañavate, J.; Pagés, P.; Saurina, J.; Colom, X.; Carrasco, F. Determination of small interactions in polymer composites by means of FTIR and DSC. *Polym. Bull.* **2000**, *44*, 293–300. [CrossRef]
27. Ahmed, F.S.; Shafy, M.; Abd El-Megeed, A.A.; Hegazi, E.M. The effect of gamma-irradiation on acrylonitrile-butadiene rubber NBR seal materials with different antioxidants. *Mater. Des.* **2012**, *36*, 823–828. [CrossRef]
28. Kazemi, H.; Mighri, F.; Park, K.W.; Fard, F.S.; Rodrigue, D. Vulcanization kinetics and properties of natural rubber hybrid composites based on maple/silica/carbon black. *Elastomery* **2019**, *4*, 227–240.
29. Schnabel, W. *Polymer Degradation: Principles and Practical Applications*; Hanser: München, Germany, 1992.
30. Zedler, Ł.; Klein, M.; Saeb, M.R.; Colom, X.; Cañavate, J.; Formela, K. Synergistic effects of bitumen plasticization and microwave treatment on short-term devulcanization of ground tire rubber. *Polymers* **2018**, *10*, 1265. [CrossRef] [PubMed]
31. Karabork, F.; Pehlivan, E.; Akdemir, A. Characterization of styrene butadiene rubber and microwave devulcanized ground tire rubber composites. *J. Polym. Eng.* **2014**, *34*, 543–554. [CrossRef]
32. Bower, D.I.; Maddams, W.F. *The Vibrational Spectroscopy of Polymers*; Cambridge University Press: Cambridge, UK, 1989.
33. Datta, S.; Antos, J.; Stocek, R. Characterisation of ground tyre rubber by using combination of FT-IR numerical parameter and DTG analysis to determine the composition of ternary rubber blend. *Polym. Test.* **2017**, *59*, 308–315. [CrossRef]

34. Hummel, D.O. *Infrared Analysis of Polymers, Resins and Additives: An Atlas. Vol. I; Plastics, Elastomers, Fibers and Resins. Part 1: Text*; Wiley: New York, NY, USA, 1971.
35. Colom, X.; Saeb, M.R.; Cañavate, J. Microstructural phenomena in ground tire rubber (GTR) devulcanized via combined thermochemomechanical and microwave processes monitored by FTIR and DTGA assisted by other techniques. *Express Polym. Lett.* **2024**, *18*, 950–961. [CrossRef]
36. Cui, H.; Yang, J.; Liu, Z. Thermogravimetric analysis of two Chinese used tires. *Thermochim. Acta* **1999**, *333*, 173–175. [CrossRef]
37. Umunakwe, R.; Umunakwe, I.J.; Odubunmi, J.O.; Malomo, D. Microstructure of CB/SIO_2 hybrid reinforced solid tyre tread compound contained Ground Tyre Rubber and their effect in fatigue resistance. *UNIOSUN J. Eng. Environ. Sci.* **2024**, *6*, 53–60.

polymers MDPI

Review

Depollution of Polymeric Leather Waste by Applying the Most Current Methods of Chromium Extraction

Ana-Maria Nicoleta Codreanu (Manea) [1,2], Daniela Simina Stefan [1,*], Lidia Kim [2] and Mircea Stefan [3,*]

1 Department of Analytical Chemistry and Environmental Engineering, Faculty of Chemical Engineering and Biotechnology, National University of Science and Technology Politehnica Bucharest, Polizu Street, No 1-7, 011061 Bucharest, Romania; anamaria.codreanu@ecoind.com
2 Department for Evaluation, Monitoring Environmental Pollution, National Research and Development Institute for Industrial Ecology, Drumul Podu Dambovitei Street, No 57-73, 060652 Bucharest, Romania; lidia.kim@incdecoind.ro
3 Pharmacy Faculty, "Titu Maiorescu" University, 22 Dâmbovnicului Street, 040441 Bucharest, Romania
* Correspondence: daniela.stefan@upb.ro (D.S.S.); stefan.mircea@incas.ro (M.S.)

Abstract: The leather industry is one of the most polluting industries in the world due to the large amounts of waste following raw hide processing but also due to the high content of chemical substances present in leather waste. The main problem with chromium-tanned leather solid waste is related to the storage, due to the ability of chromium to leach into soil or water, and also owing to the high ability of trivalent chromium to oxidize to its toxic form, hexavalent chromium. The purpose of this article is to present the most current methods of chromium extraction from solid tanned leather waste in order to obtain non-polluting leather, which can constitute secondary raw material in new industrial processes. The extraction methods identified in the present study are based on acid/basic/enzymatic hydrolysis and substitution with the help of organic chelators (organic acids and organic acid salts). In addition, this study includes a comparative analysis of the advantages and disadvantages of each identified extraction method. At the same time, this study also presents alternative chromium extraction methods based on the combination of conventional extraction methods and ultrasound-assisted extraction.

Keywords: leather solid waste; chromium removal; hydrolysis method; substitution method; circular economy

1. Introduction

Nowadays, industrial pollution is one of the biggest challenges all over the world. The leather industry is one of the most important industries in the world [1] and, at the same time, is known as one of the major polluting industries [2], due to the large amount of chemicals used, the large amounts of leather waste produced, and chemical compounds discharged into wastewater [3,4].

The leather industry is based on leather manufacturing, which is one of the oldest activities in the history of mankind [2]. This industry uses animal skins and hides as raw material [2], coming especially from the meat industry [5], to obtain a variety of products, such as belts, clothes, shoes, and bags [6]. Annually, the leather industry consumes approximately 8–9 million tons of raw skin and hide [7].

Because raw animal skins and hides are microbiologically unstable, it is necessary to apply a stabilization stage [8,9]. The resources used in the leather industry and the products resulting from the leather processing are described in Figure 1 [10].

The transformation of raw hides into a useful product called leather requires, in addition to the raw material, an energy input (process water, chemicals, process heat). This process involves pre-tanning operations, tanning operations, and post-tanning operations.

Pre-tanning operations include trimming, soaking, liming, unhairing, reliming, fleshing, deliming, bating, scudding, and pickling, while tanning operations include chrome/vegetable tanning, basification, and piling. The post-tanning operations are based on sammying, splitting, shaving, rechroming, neutralization, re-tanning, fat-liquoring, dyeing, setting, and drying. Finally, the crust leather undergoes processes of conditioning, staking, spraying, coating, toggling, trimming, biffing, measuring, plating, and polishing to obtain a finished leather product [11–14].

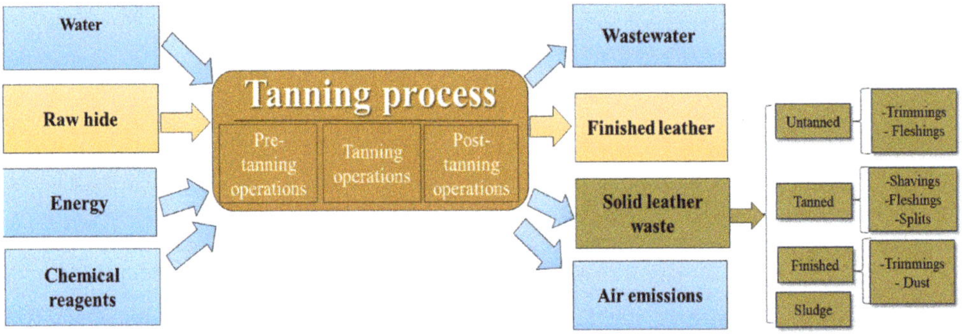

Figure 1. Tanning process: resources used and waste generated.

The tanning stage is one of the most important stages of leather processing because this process transforms raw skins into non-putrescible products [15] and protects the leather against moisture, heat, and microbial degradation, offering better dyeing characteristics and softness as well [1,2,16]. In the tanning stage, chromic or vegetable compounds can be used [11]. The chromium tanning process is the most commonly used process [17,18]. In the chromium tanning process (presented in Figure 2), chromium ions are crosslinked to the carboxylic groups by the collagen present in the leather by olation and oxolation [19].

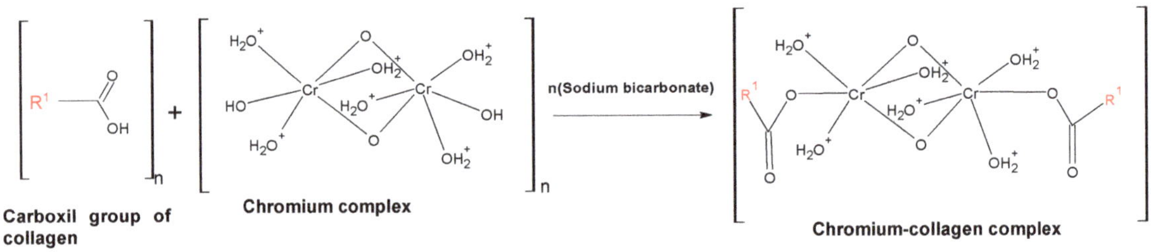

Figure 2. Description of the chrome tanning process of leather [19].

Today, over 80–90% of tanneries around the world use salts of trivalent chromium for tanning. Chromium (III) salts are among the most used tanning agents [20]. Chromium sulfate, used predominantly in the form of basic chromium sulfate ($Cr(OH)(SO_4)$), is the most used chemical compound in the leather industry for tanning and processing [20]. Also, chrome alum (chrome potassium sulfate) is another chromium-based tanning agent used in the leather industry [21]. Thus, trivalent chromium (III) is one of the most important sources of leather industry pollution due to the large amount of solid waste contaminated with the chromium produced (sludge and chromium-tanned leather waste), and the large volume of contaminated wastewater generated [20].

As a consequence of raw hide processing, solid waste and wastewater are generated besides the final product. Leather solid waste is generated during the leather production process by applying different mechanical steps, which standardize the dimensions of products. Leather solid waste can be classified as un-tanned solid leather waste (splits,

flashings, and trimmings) and organic tanned waste (shavings, splits, and trimming) [22]. During the manufacturing process of one metric ton of raw materials, the following elements are produced: 150–200 kg of finished leather, 190–350 kg of non-tanned waste, approximately 200–250 kg of tanned waste, and the rest is composed of sludge and wastewater [12,14,23,24].

Also, waste from leather processing operations is presented in the European Waste Catalog as *04 WASTES FROM THE LEATHER, FUR AND TEXTILE INDUSTRIES* [25].

The influence of the tannery process on the environment can be monitored by analyzing some parameters, such as total dissolved solids (TDS), chemical oxygen demand (COD), sulfates, chlorides, and heavy metal pollution (especially chrome pollution) [26]. Chromium present in water and soil samples can be detected by UV-VIS spectroscopy, inductively coupled plasma mass spectrometry (ICP-MS), and atomic absorption spectroscopy (AAS) [27].

Managing the large quantities of solid leather waste produced is a challenge for this industrial sector. The conventional methods of managing solid leather waste are disposal and incineration. Even if these methods are simple and inexpensive, they can only be applied to un-tanned solid waste, while for chromium-tanned leather waste, conventional methods are not practicable because they produce soil, water, and air contamination due to the hazardous reagents present in their composition. Chromium-tanned lather waste incineration can generate the conversion of Cr^{3+} to Cr^{6+}, nitrogen oxides, ammonia emissions, and hydrogen cyanide [8,22,28]. Also, the storage of solid waste from chromium-tanned leather presents the problem of chromium leaching into the soil and underground water, which can oxidize to its toxic form Cr^{6+}, affecting the health of the environment and humans. Thus, the main problem regarding the management of chromium-based solid leather waste is caused by the presence of chromium in its composition.

Chromium is an essential micro-nutrient in human and animal diets. The daily intake of chromium in the human diet is about 100 µg; it comes from cereals, fruits and vegetables, and egg yolk [29]. The recommended dose of chromium present in the diet of animals is 300 µg chromium per kg [30]. At the same time, chromium is a toxic metal for plants and microorganisms [30]. Chromium is toxic for most plants at concentrations over 100 $\mu M \cdot kg^{-1}$ dry weight [31]. This metal can exist in different oxidation states (Cr^{2+} to Cr^{6+}), but in soil, it is mostly found in two oxidation states: hexavalent chromium and trivalent chromium [32]. Chromate is the most widespread form of hexavalent chromium, and the most mobile chromium form in soil [33]. Chromium presence in the two oxidation states shows different chemical and physical characteristics, in addition to opposite toxicities: trivalent chromium presents low toxicity, while hexavalent chromium presents high toxicity [34]. Trivalent chromium's toxicity is 10 to 100 times lower than hexavalent chromium toxicity [33]. The oxidation of trivalent chromium to hexavalent chromium is pH-dependent. The possible oxidation process is described by the following equations [35]:

$$2Cr_2O_3 + 8OH^- + 3O_2 = 4CrO_4{}^{2-} + 4H_2O$$

$$2Cr_2O_3 + 2H_2O + 3O_2 = 2Cr_2O_7{}^{2-} + 4H^+$$

In 2022, hexavalent chromium was included on the Substance Priority List issued by The Agency for toxic Substances and Disease Registry, being in the top 15 [36]. Even in small doses, hexavalent chromium affects human health, causing respiratory problems, cancer, skin ulcers, and allergic reactions [17,37], and ingested in a larger dose, it causes human and animal death [38]. A large part of solid and liquid chromium-based waste comes from the leather industry, the mining industry, the dye and pigment industry, and the textile industry [33]. Even if the chromium salts used in the leather processing process do not contain hexavalent chromium, solid leather waste contains hexavalent chromium due to the dyeing additives, pigments, and fixing agents used in finishing stages, or due to non-compliance with optimal manufacturing or storage conditions.

Owing to the useful components present in solid tanned leather waste, alternative methods of managing thousands of tons of chromium solid leather waste are based on thermal decomposition (pyrolysis), microbial enzymatic degradation (aerobic degradation, anaerobic degradation, bioremediation), and obtaining value-added products [39]. Because chrome-tanned leather shavings contain approximately 90% collagen protein and between 3 and 6% chromium [40,41], which are valuable products, these can be reused in various fields by applying the circular economy principle (Figure 3) [42].

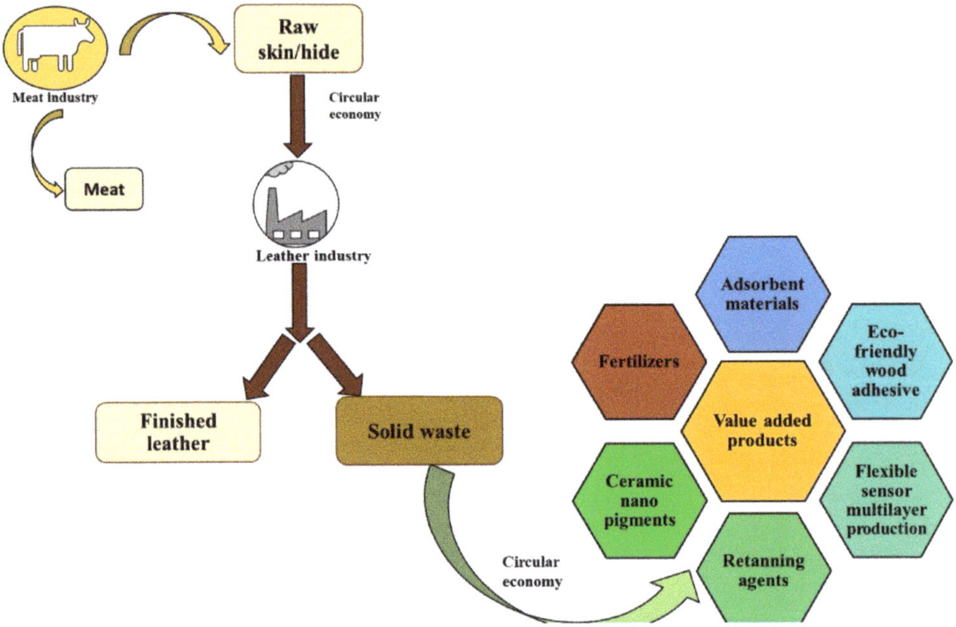

Figure 3. Valorization of solid leather waste by applying circular economy principles.

For example, due to the high content of chromium, solid waste from tanned leather can be used to produce ceramic nanopigments [43,44], or to produce re-tanning agents in the leather industry. At the same time, due to the high content of carbon and nitrogen present in the protein part, solid waste from tanned leather can be used in the production of fertilizers [45], flexible multilayer sensors [46], eco-friendly wood adhesive [47], biofuels, biochar [28,48], adsorbent materials [49,50], and construction materials [51,52].

2. Chromium Extraction Methods

The most common methods to significantly reduce the amounts of hexavalent chromium from waste, soils, and sludges are based on precipitation, reduction, adsorption, ion exchange, and reverse osmosis [33]. Tannery chromium (VI)-contaminated soils can be treated using microbial reduction methods in the presence of molasses, microbes, and native bacteria [53,54].

The recovery of chromium (III) from solid leather waste is justifiable from ecological and economical points of view, because it reduces the risk of soil, water, and air contamination. The extracted chromium can be reused, as well as other valuable components such as collagen and lipids. Chromium-tanned leather waste is the result of a covalent bond between a chromium ion and ionized carboxyl groups, belonging to aspartic and glutamic acids present in collagen [55]. Chromium–chromium bonds and chromium–collagen bonds are broken during the dechroming process of solid tanned leather waste. Dechroming can be performed by hydrolysis and by the substitution method. The most used dechroming

methods are hydrolysis (acid/basic/enzymatic) and the substitution method, with the help of organic chelators [56].

Figure 4 presents the most used methods for extracting chromium from chromium-tanned leather waste. Acidic/alkaline/enzymatic extraction can be improved by applying ultrasound-assisted extraction.

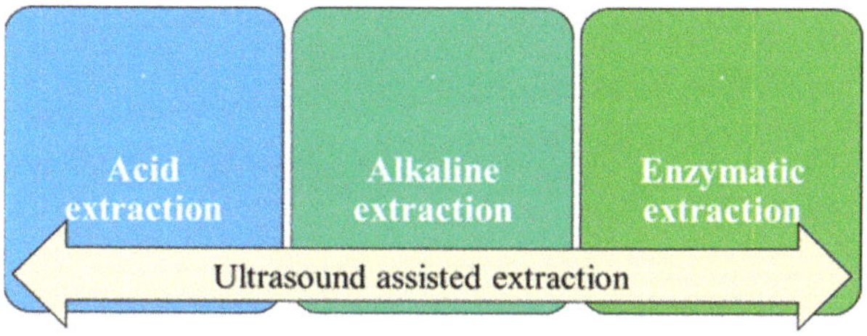

Figure 4. The most used chromium extraction methods from chromium-tanned leather waste.

Dechroming by hydrolysis is one of the most practiced dechroming methods and can be acidic, alkaline, or enzymatic. The hydrolysis method is based on the dissociation of functional groups from the Cr–collagen matrix. Acid and alkaline hydrolysis methods have one thing in common: the protein obtained (by them) presents a low molecular weight [57]. After the hydrolysis process, a protein part (peptides or amino acids) and chromium cake can be obtained, these being valuable compounds that can be reused [58].

Depending on the hydrolysis method applied, the protein part (collagen) can be slightly denatured or strongly denatured, and it can be used in various fields, depending on the degree of hydrolysis [42]. Also, collagen extracted from animal hides or from leather waste is one of the main renewable natural resources. In animal hides and skins, the most predominant type of collagen of the 29 types is type I collagen [59]. Type I collagen is a water-insoluble protein, having a triple-helix structure and therefore presenting great industrial interest [60]. Gelatin is obtained from the partial denaturation of collagen. This important biopolymer can be extracted from solid waste leather using alkaline, acid, or a combination alkaline–acid treatment, followed by thermal hydrolysis. Figure 5 presents a scheme of gelatine production [61]. The obtained gelatin can be of two types: type A gelatin (obtained in acidic conditions) and type B gelatin (obtained in alkaline conditions) [62].

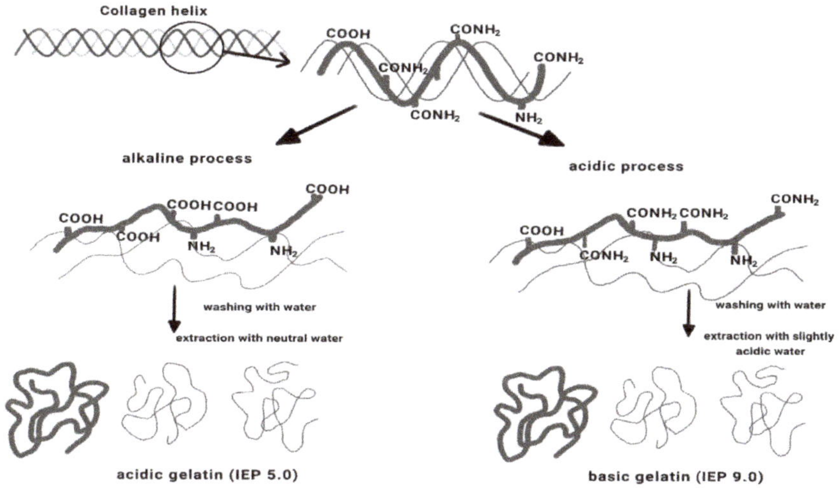

Figure 5. Scheme of acidic gelatin and basic gelatin production [61].

The hydrolysis process weakens the collagen structure, dissolves non-collagen protein, partially hydrolyzes the peptide bond, and kills bacteria. Gelatin from chromium-tanned solid leather has an excellent gel formation capacity, high biodegradability, and low cost (as shown in Table 1). The optimal extraction conditions of chromium from chrome-tanned solid waste are shown in Table 1.

Table 1. Optimal extraction conditions of chromium from chrome-tanned solid waste.

Chromium Removal Method	Chromium Removal Method	Chromium Extraction Yield	The Degree of Collagen Hydrolysis	Reference
Acid extraction	- Concentration of extraction solution = 8% H_2SO_4 - H_2SO_4:sample ratio = 11:1 - Time = 2.5 h - Temperature = 343 K	>95%	-	[63]
Acid extraction	- Concentration of extraction solution = 25 mL/L H_2SO_4 - Time = 3 or 6 days	30–60% ± 5%	3–6 ± 1%	[64]
Acid extraction	- Sample amount = 150 mg - Concentration of extraction solution = 3 mol/L HNO_3 - Temperature = 30 °C - Time = 30 min - Amplitude = 90% US bath = 37 kHz	92%	-	[65]
Acid extraction	- $H_2C_2O_4/H_2SO_4$/sample ratio = 2:1:1 - Time = 12 h - Processing times = 1 h - Stirring speed = 250 r/min - Temperature = 40 °C	95.6%	90.6%	[66]
Acid extraction	- Extraction agent: oxalic acid - Time = 36 h - Room temperature - pH = 5.5 - Cr–oxalic acid ratio = 1:3	71%	-	[67]
Acid extraction	- Concentration of potassium tartrate = 0.5 M - NaOH solution concentration = 0.25 M - Room temperature - Time = 72 h	95%	-	[19]
Acid extraction	- Concentration of sodium oxalate = 2% - Sodium oxalate/sample ratio = 200 mL/g - Thickness of sample = 0.5 mm - Temperature = 333 K - Time = 5 h - Stirring speed = 150 rpm	98%	>95%	[68]
Acid extraction	- Sample amount = 3 g - Cr^{3+}/EDTA ratio = 1:3 - Temperature = 80 °C - Time = 30 de minutes - Ultrasonic bath = 25 kHz Amplitude = 100% - 5 washing cycles with water (V = 50 mL), at a temperature of 50 °C, for 3 min each	98%	-	[69]
Acid extraction	- EDTA/Cr^{3+} ratio = 3:1 - US power = 150 W - Frequency = 20 KHz - Residence time = 60 min - Temperature = 70 °C	71.7%	-	[70]

Table 1. *Cont.*

Chromium Removal Method	Chromium Removal Method	Chromium Extraction Yield	The Degree of Collagen Hydrolysis	Reference
Alkaline extraction	- Concentration of extraction solution = 0.2 M NaOH - NaOH/sample ratio = 80 cm^3/g - Time = 1 h - Temperature = 60 °C	90%	~100%	[71]
Alkaline extraction	- Concentration of extraction solution = 0.47 M NaOH - Time = 90 min - Temperature = 70 °C	750.8 g	87.165%	[72]
Alkaline extraction	- Concentration of extraction solution = 3% NaOH - Time = 180 min - Temperature = 90 °C - NaOH/sample ratio = 5:1	~100%	-	[73]
Alkaline extraction	- Concentration of extraction solution = 4 M NaOH - NaOH/sample ratio = 0.15 - Time = 90 min - Temperature = 423 K	85%	98%	[74]
Alkaline extraction	- H_2SO_4 concentration = 0.1 N H_2SO_4 - Dose of Gamma radiation ^{60}Co = 60 Krad - Concentration of extraction solution = 1 N NaOH	~100%	25–40%	[75]
Alkaline extraction	- Concentration of extraction solution 1 = 2 g/L NaOH, stirring for 30 min at 30 °C, urea concentration = 40 g/L - Concentration of extraction solution 2 = 50 g/L H_2SO_4, stirring for 1 h at 30 °C - Concentration of extraction solution 3 = 40 g/L CaOH, stirring for 2 ore at 30 °C - Concentration of extraction solution 4 = 50 g/L H_2SO_4, stirring for 1 h at 30 °C	97%	10%	[76]
Enzymatic extraction	- Extraction solution concentration = 6% MgO - Bating enzyme concentration = 0.75% - Time = 30 h - Temperature = 33–37 °C - pH = 8.3–8.5	99.99%	-	[77]
Enzymatic extraction	- Extraction solution MgO - Stirring speed = 60 rpm - Temperature = 70 °C - Time = 6 h - *Bacillus subtilis enzyme A proteolytic* activity = 130.5 U/mL - pH = 9 - Time = 15 h - Temperature = 45 °C - Stirring speed = 60 rpm	~100%	-	[78]
Enzymatic extraction	- Extraction solution concentration = 3% MgO - Extraction solution concentration = 3% CaO - Temperature = 80 °C - Time = 4 h - 1398 neutral protease concentration = 0.125% - Temperature = 46 °C	~100%	>60%	[79]

3. Acid Extraction of Chromium

Acid methods used inorganic acids (sulfuric acid, hydrochloric acid, and azotic acid) and organic acids (oxalic acid, citric acid, tartric acid) for the dechroming of chrome-tanned solid waste [66]. Figure 6 presents the extraction of chromium from chromium-tanned leather waste with the help of an inorganic acid and an organic acid [1]. The extraction of

chromium with inorganic acids, based on the acid hydrolysis method, destroys the collagen matrix while the extraction of chromium with organic chelators, based on the substitution method, keeps the collagen matrix intact.

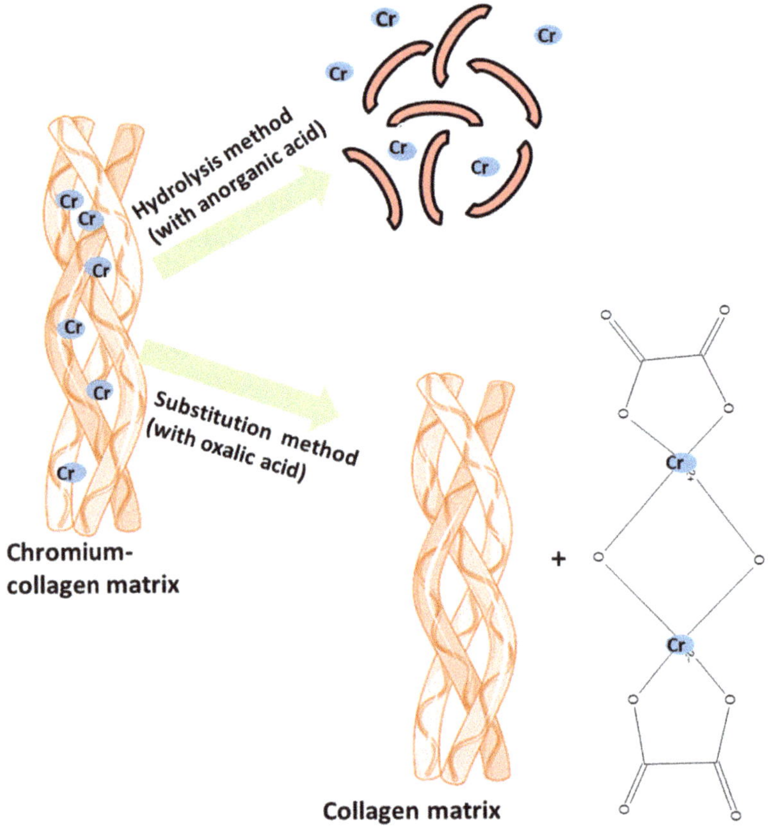

Figure 6. Extraction of chromium by hydrolysis method and substitution method.

In this method, the trivalent chromium, present in the chromium–collagen matrix, combines with the strong anionic groups of the acid, and thus, a soluble complex that can be extracted from the solution is obtained [63]. Conventionally, oxalic acid has been the best acid reagent for dechroming, followed by sulfuric acid, but the large-scale use of organic acids in the treatment of tanned leather solids is limited because of the high prices. Consequently, sulfuric acid became the first choice of many researchers, being used in numerous studies as a dechroming agent for tanned leather [63]. The main advantage of acid methods is the possibility of reusing acid solutions containing chromium in the leather industry in the tanning, pickling, and dyeing stages [66].

In acid hydrolysis, inorganic acids H^+ can replace trivalent chromium ions and are combined with the COO^- of collagen. The advantages of dechroming using the acid method are stable products, low cost, and easy control and implementation [66].

The dechroming process of chromium-tanned leather samples using sulfuric acid at a low temperature was studied by varying some indicators such as sample dimensions, temperature, extraction time, sulfuric acid concentration, and sodium sulfate concentration [64]. By increasing the concentration of sulfuric acid from 35 mL/L to 200 mL/L, an improvement in the chromium extraction performance could be observed, but a high concentration of sulfuric acid led to matrix degradation. The temperature also has a similar

influence on the extraction process; by changing the temperature from 296 K to 313 K, the chrome recovery yield improves, but the degradation of the protein matrix takes place. The variation in the size of the leather waste (4 mm or at sizes of 5 cm × 3 cm) did not significantly influence the chromium extraction process; the same insignificant influence was also observed in the case of varying the L-S ratio (from 5:1 to 10:1). The research team proposed two optimal chromium extraction formulas. If a high chrome recovery yield (55–60 ± 5%) is desired, but with a high degree of denaturation of the tanned leather (dissolved TOC is 3–6%), the leather waste can be subjected to a solution treatment of H_2SO_4 25 mL/L, in a L-S ratio between 5:1 and 10:1, for 3–6 days at 313 K. If a moderate chromium recovery yield is desired (35 ± 5%), but with a low degree of denaturation of tanned leather (TOC < 3–6%), the leather waste can be treated with 25 mL/L H_2SO_4 solution, in a L-S ratio between 5:1 and 10:1, for 6 days at 296 K [64].

In addition, Wang et al. studied the complete removal of chromium from leather waste in two stages. The first stage was based on the use of the acid extraction method of chromium. The second stage of ion exchange aimed to remove the chromium found in the filtrate collected in the first stage with the help of ion exchange resins. By varying some parameters involved in the extraction process, such as the liquid–solid ratio, sulfuric acid concentration, contact time, and reaction temperature, they could observe that the factor that most strongly influences the extraction yield is the reaction temperature (50–70 °C), followed by the sulfuric acid concentration (6–10%), the liquid–solid ratio (9:1 mL/g to 11:1 mL/g), and finally, the reaction time (2.5 h to 3.5 h). By setting the liquid–solid ratio at 11:1 mL/g, the sulfuric acid concentration at 8%, the contact time at 2.5 h, and the reaction temperature at 70 °C, an efficient removal of chromium from leather waste was obtained, close to 100% [63].

However, conventional acid/base/enzymatic hydrolysis can break the peptide bonds in collagen, thus resulting in destroyed collagen, which can be difficult to remove from the hydrolysate. Chromium extraction from chromium-tanned leather waste by means of organic acids or their salts has the advantage of efficiently removing chromium without significantly degrading the collagen matrix.

Dechroming using organic acids/salts is based on organic acid anion coordination with trivalent chromium using the substitution method (described in Figure 7). The dechroming process presents two stages. In the first stage, the breaking of the chromium–collagen bonds present in the solid tanned leather waste takes place through the substitution reaction (exchange) of a water molecule around the first chromium atom with the ionized carboxylic group belonging to the organic chelator, followed by the breaking of the first bond in the chromium–collagen to restore the coordination number of chromium. In the second stage, the ionized carboxylic groups of the organic chelator attack the second chromium atom, followed by the breaking of Cr–collagen bonds, to restore the coordination number [19].

The main advantages of this method are the low degree of collagen hydrolysis, high collagen recovery yields, and low amounts of dangerous materials generated [66]. At the same time, the loss of collagen from chrome-tanned solid waste using organic acid extraction is lower than alkaline methods [66].

The efficiency of chromium separation from chromium leather waste using organic acids (acetic acid and citric acid) and organic salts (potassium oxalate and potassium tartrate) was tested by Malek and colleagues [19]. The experimental results demonstrated a direct influence of some parameters involved in the process, such as the type of chelating agent used, the type of extraction medium, the concentration of the chelating agent, the concentration of the extraction medium, the temperature, and the contact time, on the chromium extraction yield (III). The tests regarding the choice of the organic chelator carried out in the neutral environment demonstrated the fact that the salts of organic acids have a higher capacity to extract chromium, compared to the organic acids used in the extraction processes. This fact is due to the presence of the alkaline environment in the case of salts, which can form stronger bonds between the carboxylic groups and chromium more easily, while the acidic environment can inhibit the reaction. Also, by testing the

medium suitable for the extraction process, it was possible to conclude that, for all four organic chelating agents, the highest extraction yields were obtained in the basic medium (94.3% when potassium tartrate was used in the basic medium). By studying the effect of the potassium tartrate concentration on the extraction yield, it was observed that, by increasing the concentration of the chelating agent (0.005 M, 0025 M, 0.05 M, 0.25 M, 0.5 M, 1 M), the pH values also increased. The stabilization of the pH leads to a significant increase in the extraction yield; thus, the optimal value of the potassium tartrate concentration was 0.5 M. Also, by progressively increasing the concentration of the alkaline solution (0.01 M, 0.05 M, 0.1 M, 0.25 M, 0.5 M, 1 M), the pH tends to increase, and the Cr removal efficiency tends to follow the same trend. A concentration higher than 0.5 M leads to the degradation of the sample matrix, due to the strong manifestation of the phenomenon of hydrolysis of hydroxyl groups on the chromium–collagen complex. Through the simultaneous study of the influence of temperature (25, 30, and 40 °C) and contact time (30, 90, and 180 min), the results obtained showed that temperature and reaction time have significant effects on the efficiency of chromium extraction. A decrease in the residual chromium concentration could be observed by keeping the temperature constant at 40 °C and varying the contact time.

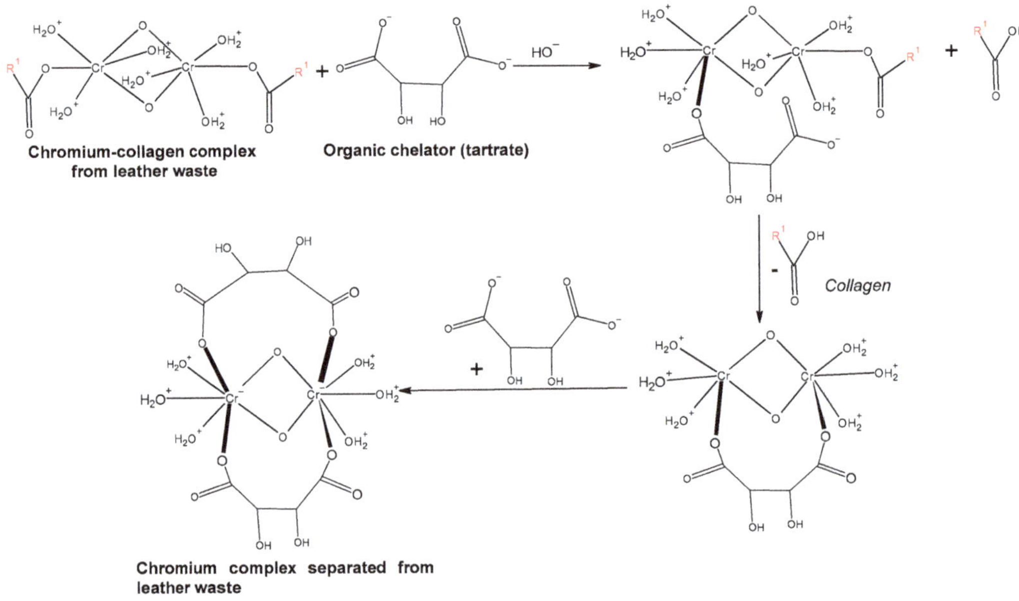

Figure 7. Description of leather waste dechroming process using the substitution method [19].

Also, the extraction efficiency of chromium (III) from the solid tanned leather waste with the help of organic acid salts (sodium gluconate, trisodium citrate, EDTA-2Na, sodium oxalate, sodium, and potassium tartrate) was also studied [68]. The studies about the choice of chelating agent demonstrated that all organic chelators tested show good chromium extraction capabilities, due to the presence of carboxylic groups in their molecules, which show a strong ability to complex chromium. The chromium extraction efficiency decreases with increasing chelating agent molecule size and with an increase in the carboxylic group number. The most suitable chelating agent was sodium oxalate. The existence of a directly proportional relationship between the extraction yield and the sodium oxalate concentration (1–3%) was observed. The optimal value of sodium oxalate concentration is 2%, because after this value, the phenomenon of denaturation of the collagen matrix appears. Also, the extraction yield of chromium is significantly influenced by the increase in temperature (318–338 K), and it increases with an increase in temperature, with the optimal temperature being 333 K, because above this temperature, the collagen matrix is destroyed. Further

studies on the influence of the thickness of tanned leather waste samples (0.5–2.3 mm) and the stirring speed (0 to 150 rpm) demonstrated that decreasing the thickness of the waste improves the extraction yield, and increasing the speed of stirring also leads to a significant improvement of the extraction yield [68].

In addition, the influence of heteroatomic organic compounds (8-hydroxyquinoline, acetylacetone, dithizone, thiourea DMSO) and organic compounds containing a carboxyl group in the molecule (maleic acid, citric acid, oxalic acid, tartaric acid, salicylic acid, EDTA) on the dechroming process of tanned leather solid waste was studied [67]. The experimental results demonstrated that the most effective chelating agents are those containing the carboxyl group in the molecule, and the best chromium extraction yield from these wastes was obtained for oxalic acid (71%), followed by citric acid (63%) and tartaric acid (62%), the extractions being carried out in a single step. In contrast, the compounds containing sulfur atoms in the molecule (dithizone, thiourea, DMSO) showed the weakest results (0–3%). Studies on the influence of pH, temperature, and contact time on the extraction efficiency of oxalic acid and citric acid demonstrated that an increase in pH towards the neutral medium significantly improved the extraction yield for both acids, and using higher temperatures significantly decreased the extraction time and the chromium extraction yield. At temperatures exceeding 60 °C, the protein matrix in the analyzed waste was destroyed.

Considering the advantages and disadvantages of the two methods of extraction in an acidic environment (extraction with inorganic acids and extraction with organic acids), the team coordinated by Tian studied the extraction of chromium by applying a combined acid extraction method, using a mix between an organic acid (oxalic acid) and an inorganic acid (sulfuric acid). The studies carried out followed the influence of the mass ratio between oxalic acid, sulfuric acid, and tanned leather waste but also the influence of the time interval at which the acid solution was replaced on the chrome extraction yield and on the collagen recovery yield. By varying the mass ratio of oxalic acid/sulfuric acid/leather waste in the range of 1:0.5:1–3:1.5:1, it was possible to observe that the optimal ratio is 2:1:1, as it obtained a degree of dechroming of approximately 96.5% and a collagen recovery yield of 90.6%. Also, by varying the time interval in which the acid solution was replenished, the best collagen recovery yield and the best degree of dechroming over 90% were obtained, and the solution changed 3 times in 12 h. The main advantages of the developed method included obtaining collagen with a high molecular weight, a high yield of chromium extraction, and moderate costs [66].

4. Alkaline Extraction of Chromium

Alkaline hydrolyses can be carried out using calcium oxide, magnesium oxide, sodium hydroxide, potassium hydroxide, and calcium hydroxide. In alkaline hydrolysis, the coordination of the chromium ion (III) to the hydroxide ion takes place, and the precipitation of trivalent chromium in $Cr(OH)_3$ can be observed, while the collagen is hydrolyzed (polypeptide chains break into small peptide fragments or amino acids). Cr^{3+} formed from the soluble hydroxy complex can later be separated by extraction with another solvent (for example, quaternary ammonium compounds) [71]. The main advantages of alkaline extraction are the ease of solid–liquid mixture separation and the lack of high technology and high costs, while the disadvantages include obtaining hydrolyzed collagen with a low molecular weight and a recovery yield of protein that does not exceed 50% [66].

The study of the influence of the sodium hydroxide concentration, contact time, and temperature on the chromium (III) extraction process from the composition of tanned leather waste through the basic hydrolysis process was carried out by Wionczyk and colleagues [71]. The experimental results demonstrated that the extraction yield of chromium from the chromium–collagen matrix can be improved by increasing the concentration of sodium hydroxide from 0.2 M to 0.3 and increasing the degree of hydrolysis of the protein matrix at the same time. By increasing the temperature in the range of 313–343 K, it could be observed that a high chromium extraction yield was reached in a shorter time. Under optimal conditions of 60 °C, one-hour contact time, and 0.2 M NaOH concentration,

chromium (III) was efficiently separated from the collagen–protein matrix, with the extraction yield reaching the value of 90%. However, following the application of the optimal extraction conditions, the collagen present in the solid waste from the skin was completely hydrolyzed, this aspect constituting a serious disadvantage [71].

Studies on the influence of the hydrolysis time (1–3 h) and the concentration of the extraction agent (1–3% NaOH) at a temperature of 90 °C on the results of the basic hydrolysis process of leather shaving scraps were also carried out by Pahlawan et al. [73]. The experimental results demonstrated that, by increasing the contact time to 3 h and the NaOH concentration to 3%, a hydrolysate with a protein content of 6.64% and a chromium content of 47.55 ppm could be obtained (the initial chromium content in leather shaving scraps being 23,176.14 ppm) [73]. Similar studies regarding the influence of the hydrolysis time, as well as the concentration of sodium hydroxide, on the extraction process of chromium from chrome shavings were also carried out by the research team led by Tahiri [80]. The optimal conditions were a 15 min reaction time and 0.5 M concentration of sodium hydroxide solution [80].

The alkaline extraction of chromium from chrome leather shavings with the help of sodium hydroxide was also studied by Barra Hinojosa and collaborators [72]. In the study, the relationship between working conditions (contact time and sodium hydroxide concentration) and the recovery yields of hydrolyzed collagen and the recovery yields of chromium, was studied. It was found that by increasing the concentration of sodium hydroxide (0.1–0.5 M), the hydrolysis reaction was improved, and the optimal concentration was 0.47 M. By varying the contact time between 30 min and 120 min, it was found that by increasing the contact time, the separation of chromium from the chromium–collagen matrix is improved, and the optimal contact time was 90 min. Thus, in optimal conditions (70 °C, 0.47 M NaOH, 90 min), 45 L of collagen hydrolysate with 87.16% protein collagen and 1.17% residual chromium were recovered [72].

Also, the research team of Ferreira et al. studied the chromium extraction efficiency from solid salts using sodium hydroxide as the extraction medium, and also the effects of temperature, contact time, and S-L ratio [74]. Studies of the influence of the sodium hydroxide solution concentration (0–4 M) demonstrated that the NaOH concentration is one of the most important factors in the chromium extraction process. By increasing the NaOH concentration, as in the case of the study carried out by Wionczyk, an increase in the chromium extraction yield was observed, as well as significant degradation of the collagen matrix. At the same time, studies have shown that temperature (373–473 K) and contact time (1–24 h) are important factors. Their increase leads to significant degradation of the collagen matrix (almost completely). In contrast, the S-L ratio has no significant effect on the chromium (III) extraction yield [74]. The optimal conditions were a temperature of 423 K for 1.5 h with NaOH 4 mol/L solution, and a solid-to-liquid (S/L) ratio (*w*/*w*) of 0.15 or 0.2.

In order to reduce the disadvantage of the destruction of the collagen matrix during the process of chromium extraction from tanned leather waste, the research team led by Poulopoulou implemented two additional steps to protect the protein matrix with H_2SO_4 and Gamma radiation, in addition to the extraction step with NaOH [75]. Also, the research carried out in the study demonstrated that the introduction of a final extraction step with hydrogen peroxide led to the improvement of the chromium extraction yield, with the final amount of chromium in the tanned leather solid waste being below 10 ppm. The study of the influence of the concentration of sulfuric acid demonstrated that an increase in the concentration of sulfuric acid (0.1–1 N) directly proportionally influences the degree of denaturation of the protein, as well as the chromium extraction yield, while the study of the influence of the Gamma radiation dose demonstrated that an increase in this parameter leads to an increase in the denaturation of the protein matrix. A concentration of NaOH higher than 1N leads to the destruction of the protein part, and a concentration below this value significantly reduces the chromium extraction yield [75].

Even if the extraction yields are high when alkaline extraction is used, in order to reduce the disadvantage given by the high degree of collagen hydrolysis, this method must be combined with acid hydrolysis. Thus, to improve the dechroming yield, but also to keep a low degree of collagen hydrolysis, an alternating acid–base extraction process was tried on solid waste from tanned leather [76]. Chromium extraction was carried out in four stages, as follows: basic extraction with NaOH; acid extraction with H_2SO_4; basic extraction with $Ca(OH)_2$; and acid extraction with H_2SO_4. At the same time, to improve the results obtained in the study, the influence of hydrolysis assistants (ammonium hydroxide, diethanolamine, methyl-urea, guanidine hydrochloride, and urea), reaction temperature, sulfuric acid concentration, and calcium hydroxide concentration were studied. It could be observed that all the compounds with amino groups used as hydrolysis assistants were able to increase the degree of dechroming, with urea having the best capacity (65.66%). It could be observed that the largest amounts of chromium were extracted in the first two stages. By varying the concentration of $Ca(OH)_2$, it was possible to observe that a high degree of chromium extraction was obtained at the optimal concentration of 40 g/L; below this value, the degree of collagen hydrolysis was lower. It has been reported that Ca^{2+} has a compressive effect on collagen fibrils at high concentrations and thereby protects the collagen fibrils from excessive damage [76].

5. Enzymatic Extraction of Chromium

Enzymatic extraction of chromium is based on the hydrolysis of collagen and chromium release at the same time [66]. Enzymatic hydrolysis can use different enzymes, such as crude proteolytic enzyme extracts obtained from the cultivation of new Bacillus subtilis strains, bating enzyme, and 1398 neutral protease. The advantages and disadvantages of this method are presented in Figure 8 [66].

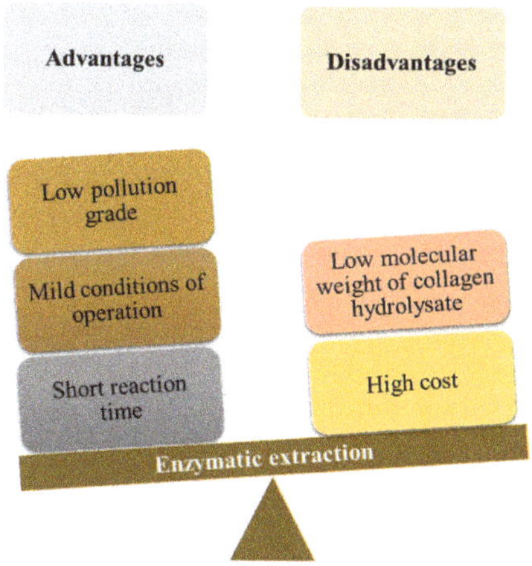

Figure 8. The main advantages and disadvantages of the enzyme extraction of chromium from chromium-tanned solid wastes.

The molecular weight of the collagen hydrolysate can be influenced by the reaction mixture and concentration of enzymes. Also, the economy of the process can be influenced by a whole series of factors, such as enzyme price, enzyme concentration, reaction speed, reaction time, and reaction yield [81].

Enzymatic hydrolysis does not take place individually, but together with chemical hydrolysis. Due to the limitation given by the recovery of the protein part from the obtained hydrolysate and the chromium recovery, the combination of two types of hydrolysis was tried. Enzymatic hydrolysis is performed in two steps: a chemical pre-treatment step, followed by the enzyme addition step. A pre-treatment step is mandatory in enzymatic

hydrolysis because gelatin produced after the first step helps in obtaining hydrolyzed collagen in the second stage by means of enzymes [58].

For example, alkaline-enzymatic hydrolysis is composed of two stages: in the first stage, basic hydrolysis takes place (with the help of NaOH, CaO, MgO), and in the second stage, enzymatic hydrolysis is applied to the residue left from the previous stage.

Asava and collaborators tried to apply a combined alkali–enzyme hydrolysis process, to recover chrome and to improve the recovery efficiency of the protein part from tannery chrome shavings. They applied a stage of pre-treatment of the waste with MgO to denature, degrade, and increase the vulnerability of the collagen to the proteolytic attack, which followed in the next stage with the help of bating enzymes. After the first stage of basic hydrolysis with MgO 6%, the hydrolysis yield was 58.2% and the chromium content in the hydrolysate was 2.89 ppm (initial content was 3.04%). In the enzymatic hydrolysis stage, the effects of enzyme concentration (0.25–1%) and contact time (0–40 h) on protein recovery yield were studied. It could be observed that at a 0.75% enzyme concentration and 30 h contact time, the total yield of protein recovery exceeded 79.45%, and chromium was extracted in a proportion of 99.99% [77].

The extraction of chromium and hydrolyzed collagen from solid chromium-tanned waste combining two methods (alkaline hydrolysis and enzymatic hydrolysis) was also studied by Dettmer and colleagues [78]. In the alkaline hydrolysis stage, they studied the influence of two extraction reagents (NaOH 0.1 M and MgO) on the efficiency of the chromium and collagen protein extraction process. The experimental results demonstrated that, in the same reaction conditions (stirring of 60 rpm, 70 °C, and 15 h), a larger amount of protein was extracted by using MgO and 0.133 mg/l total chromium. In the enzymatic hydrolysis stage, the influence of two crude proteolytic enzymes, enzyme A (obtained by cultivating *Bacillus subtilis Blbc* 11), and enzyme B (obtained by cultivating *Bacillus subtilis Blbc* 17), on the chromium extraction yield and the hydrolysate extraction yield from chromium cake obtained in the first stage was studied [78]. It can be observed that the liquid extracted from the hydrolysis carried out by enzyme A had a lower chromium content (0.171 mg/ L), while enzyme B had a higher chromium content (0.359 mg/L) [78].

In addition, the use of two oxide mixes (CaO and MgO) in the alkaline hydrolysis step, followed by enzymatic hydrolysis with the help of 1398 neutral protease, to recover the hydrolyzed protein and chromium from chrome-tanned pig leather shavings was suggested by Qiang et al. [79]. Studies on the influence of MgO/CaO alkali dosage (0:1, 2;3, 1:1, 3:2, 1:0), alkaline hydrolysis temperature (60–100 °C), alkaline reaction time (3–7 h), enzyme dosage (0.03–0.15%), and enzyme hydrolysis temperature (42–50 °C) on the hydrolyzed protein extraction yield and on the chromium extraction yield were performed. The results showed that a hydrolyzed protein recovery yield of more than 60% was obtained when using a dosage of calcium oxide and magnesium oxide of 3%, an alkaline hydrolysis temperature of 80 °C, a time of 4 h, an enzyme dosage of 0.125%, and an enzyme hydrolysis temperature of 46 °C, while the chromium content in the final hydrolysate after desalination was less than 50 mg/Kg [79].

Moreover, the research team led by Rahaman carried out parallel studies regarding the influence of acid hydrolysis, the influence of acid–alkaline hydrolysis, and the influence of acid–alkaline–enzymatic hydrolysis on the chrome extraction yield from chrome shaving dust [82]. In the studies on acid hydrolysis, the influence of the type of acid extraction agent (H_2SO_4, HNO_3, HCl) and their concentrations (1–6 M) was studied. Experimental studies demonstrated that HNO_3 6 M has the best chromium recovery capacity (70.09%). The studies on acid–alkaline hydrolysis with the help of CaO, MgO, and NaOH, and the acids H_2SO_4, HNO_3, and HCl, demonstrated that all three alkaline media had nearly the same ability to recover chromium, but the mixture of MgO and HNO_3 presented the best extraction capacity of chromium (43.71%). Following the studies on acid–alkaline–enzymatic hydrolysis carried out with the help of the three acids, the three bases listed previously, and proteolytic enzymes, it was concluded that this technique significantly improves the

chromium extraction yield. The best chromium extraction yield was 55.51% when MgO, NHO_3, and proteolytic enzymes were used.

6. Ultrasound-Assisted Extraction of Chromium

Chromium extraction processes that are based on enzymatic treatment or alkaline–acid treatments are strongly affected by the formation of metabolic inhibitory products and by the high costs of purchasing pure enzymes in the case of enzymatic hydrolysis, and by the use of concentrated reagents, numerous steps, and a relatively long reaction time (72 h) in the case of acid–alkaline treatments. In addition, both processes lead to significant degradation of the protein matrix in the hide waste, reducing the quality of the raw material. Considering the disadvantages listed above, ultrasound-assisted extraction of chromium can be a sustainable alternative to existing methods for the extraction of chromium from leather waste. Ultrasound-assisted extraction can be used as an additional method to the classic extraction methods (basic, enzymatic, acid hydrolysis, and substitution with organic chelators) in order to improve the extraction yield. Studies have shown that the additional application of ultrasound in the chromium extraction process from solid tanned leather waste significantly improves the chrome extraction yield from this waste, due to the effects caused by acoustic cavitation [69].

The ultrasound propagation process involves cavitation bubbles, vibrations, crushing, mixing, and other comprehensive effects in the liquid medium [83]. When ultrasound is applied in extraction processes in a liquid medium containing solids, at the liquid–solid interface, bubbles can be formed with further collapse (cavitation), increasing the surface area of the substrate, as well as the transport rates of extraction solution [84]. The main advantages of ultrasound-assisted extraction include a short extraction time but also the use of mild extraction conditions (temperature, concentration, and amounts of reagents) [85].

The results of the study carried out by Bizzi and colleagues highlighted the fact that, by the additional application of an ultrasonic stage in the chromium extraction process in an acidic medium (HCl, NHO_3, H_2SO_4, CH_2O_2, $C_2H_2O_4$), the chromium extraction yield was significantly improved [65]. Studies revealed that ultrasound-assisted extraction leads to a chromium extraction yield of 92%, while the application of the agitation step leads to a chromium extraction yield of 65%. Also, the studies related to the influence of some operating parameters, such as the type of extraction solution, the concentration of the extraction agent, the temperature, the time, the amount of the sample, and the amplitude, demonstrated the existence of a direct relationship between the extraction yield of chromium (III) and these parameters. By testing five inorganic acids used as chromium extraction media (HCl, NHO_3, H_2SO_4, CH_2O_2, $C_2H_2O_4$), it was revealed that nitric acid had the highest chromium extraction capacity, owing to its oxidation power and demineralization potential, followed by sulfuric acid. By increasing the concentration of nitric acid (0.1–4 mol/L), a significant increase in chromium extraction yield could be observed because the high proton concentrations of HNO_3 destabilized the strong Cr-Cr bonds between neighboring chromium atoms in the chromium-collagen matrix, but after a certain concentration, a destabilization of the protein structure occurred. The study of the parameters directly involved in the ultrasound stage demonstrated that there is an increase in the chromium extraction yield at low frequencies (25–37 KHz) using the ultrasound bath, but also an increase in extraction with an increase in the applied amplitude (10–90%). Studies on the influence of temperature (10–90 °C) have shown that there is an improvement in the extraction yield with increasing temperature, but above 30 °C, degradation of the protein matrix occurs, as it is well known that collagen denaturation occurs at 37 °C [65].

The influence of the additional application of the ultrasound process to the chemical extraction process of chromium with EDTA was studied [69]. The results demonstrated that by applying ultrasound-assisted extraction, the total extraction time (including extraction and washing steps) of chromium found in the waste composition was reduced to 45 min. In addition, it was possible to conclude that the extraction yield is significantly influenced

by the variation of other parameters involved in the extraction process of chromium from tanned leather waste (extraction temperature, L-S ratio, extraction time). The study of the influence of the S:L ratio (Cr^{3+}: EDTA) showed that the extraction yield improves significantly when the ratio increases (in the range of 1:0–1:6), but the optimal value is 1:3, because in the case of a ratio of 1:6, a similar yield is obtained. The chromium extraction yield follows a similar trend: when the extraction temperature is varied (60 °C to 90 °C), but at temperatures higher than 80 °C, a degradation of the skin waste is observed, while the variation of the ultrasonication time (30–120 min) does not significantly influence the yield [69].

At the same time, a similar study on the influence of ultrasound in the presence of an organic acid salt (EDTA) on the extraction yield of Cr (III) was carried out by Pedrotti and collaborators [70]. By studying the influence of the ultrasonic power, the applied temperature, and the contact time, it was possible to observe a direct influence between increasing the values of these parameters and the amount of chromium extracted from the analyzed leather waste. Increasing the temperature (30–80 °C) led to an increase in the extraction yield, but at temperatures higher than 70 °C, a degradation of the solid skin waste was observed, turning it into a gelatinous fraction. Also, a decrease in contact time and non-breakage temperature to achieve high extraction yields was observed when the ultrasonic power increased. By varying the washing time (3–10 min), an insignificant influence on the extraction yield was observed [70].

7. Conclusions

The purpose of this review was to present the most current methods used for dechroming solid chrome-tanned leather, in order to reuse it as an alternative raw material in new industrial sectors by applying the principles of circular economy.

In this study, the latest acid, basic, and enzymatic extraction techniques for trivalent chromium were identified based on the hydrolysis method or on the substitution method, and at the same time, the advantages and disadvantages of each identified method were also discussed. It could be observed that by using the hydrolysis method to extract chromium from tanned leather waste, chromium recovery yields between 30 and 100% can be obtained, but at the same time, the collagen matrix can be partially (3–10%) or totally (~100%) destroyed. On the other hand, using the substitution method with the help of organic chelators, high chromium extraction yields (over 71%) were obtained without significantly degrading the collagen matrix, but the purchase costs of the organic reagents used were relatively higher. In addition, it could be observed that by applying the ultrasonic process in chromium extraction experiments, the extraction time, concentration, and amount of reagents were significantly reduced, while the chromium extraction yields exceeded 71.7%.

Based on the literature studies carried out, it is possible to conclude that a combination of the hydrolysis and substitution methods with the ultrasound process leads to significantly improved chromium removal yields from leather solid waste and a low collagen matrix degradation yield.

Author Contributions: Conceptualization, D.S.S. and A.-M.N.C.; methodology, L.K. and M.S.; validation, D.S.S., L.K. and M.S.; investigation, A.-M.N.C.; resources, D.S.S.; data curation, M.S.; writing—original draft preparation, A.-M.N.C.; writing—review and editing, D.S.S. and M.S.; visualization, L.K.; supervision, D.S.S. All authors have read and agreed to the published version of the manuscript.

Funding: This research was funded by Ministry of Research, Innovation and Digitalization of Romania, through the Nucleu Programme through contract 3N/2022, Project PN 23 22 04 01.

Conflicts of Interest: The authors declare no conflicts of interest.

References

1. Wang, L.; Chen, M.; Li, J.; Jin, Y. A novel substitution-based method for effective leaching of chromium A novel substitution-based method for effective leaching of chromium kinetics and mechanism studies. *Waste Manag.* **2022**, *103*, 276–284. [CrossRef] [PubMed]
2. Satish, M.; Madhan, B.; Sreeram, K.J.; Rao, J.; Nair, B.U. Alternative carrier medium for sustainable leather manufacturing—A review and perspective. *J. Clean. Prod.* **2016**, *112*, 49–58. [CrossRef]
3. Bulijan, J.; Reich, G.; Ludvik, J. Mass Balance in Leather Processing. United Nations Industrial Development Organisation (UNIDO). 9 August 2000. Available online: https://www.unido.org/sites/default/files/2009-05/Mass_balance_in_leather_processing_0.pdf (accessed on 20 December 2023).
4. Tzoumani, I.; Lainioti, G.C.; Aletras, A.; Zainescu, G.; Stefa, S.; Meghea, A.; Kallitsis, J. Modification of Collagen Derivatives with Water-Soluble Polymers for the Development of Cross-Linked Hydrogels for Controlled Release. *Materials* **2019**, *12*, 4067. [CrossRef]
5. Rahmawati, D.; Setyadewi, N.M.; Sugihartono. Extraction and characterization of gelatin from skin trimming pickled waste of tannery. *Earth Environ. Sci.* **2019**, *306*, 012022. [CrossRef]
6. Rigueto, C.V.T.; Nazari, M.T.; Rosseto, M.; Massuda, L.A.; Alessandretti, I.A.; Piccin, J.S.; Gettmer, A. Emerging contaminants adsorption by beads from chromium (III) tanned leather waste recovered gelatin. *J. Mol. Liq.* **2021**, *330*, 115638. [CrossRef]
7. Ayele, M.; Limeneh, D.I.; Tesfaye, T.; Mengie, W.; Abuhay, A.; Haile, A.; Gebino, G. A Review on Utilization Routes of the Leather Industry Biomass. *Adv. Mater. Sci. Eng.* **2021**, *2021*, 1503524. [CrossRef]
8. Rosu, L.; Varganici, C.D.; Crudu, A.M. Ecofriendly wet-white leather vs. conventional tanned wet-blue. *A Photochem. Approach. J. Clean. Prod.* **2018**, *177*, 708–720. [CrossRef]
9. Alibardi, L.; Cossu, R. Pre-treatment of tannery sludge for sustainable landfilling. *Waste Manag.* **2016**, *52*, 202–211. [CrossRef] [PubMed]
10. Famielec, S. Chromium Concentrate Recovery from Solid Tannery Waste in a Thermal Process. *Materials* **2020**, *13*, 1533. [CrossRef]
11. Kanagaraj, J.; Senthilvelan, T.; Panda, R.C.; Kavitha, S. Eco-friendly waste management strategies for greener environment towards sustainable development in leather industry: A comprehensive review. *J. Clean. Prod.* **2015**, *89*, 1–17. [CrossRef]
12. Mella, B.; Glanert, A.C.; Gutterres, M. Removal of chromium from tanning wastewater and its reuse. *Process Saf. Environ. Prot.* **2015**, *95*, 195–201. [CrossRef]
13. Hu, J.; Xiao, Z.; Zhou, R.; Deng, W. Ecological utilization of leather tannery waste with circular economy model. *J. Clean. Prod.* **2011**, *19*, 221–228. [CrossRef]
14. Dixit, S.; Yadav, A.; Dwivedi, P.; Das, M. Toxic hazards of leather industry and technologies to combat threat: A review. *J. Clean. Prod.* **2015**, *87*, 39–49. [CrossRef]
15. Sundar, V.J.; Gnanamani, A.; Muralidharan, C.; Chandrababu, N.K. Recovery and utilization of proteinous wastes of leather making: A review. *Rev. Environ. Sci. Biotechnol.* **2011**, *10*, 151–163. [CrossRef]
16. Kanagaraj, J.; Chandra Babu, N.K.; Mandal, A.B. Recovery and reuse of chromium from chrome tanning waste water aiming towards zero discharge of pollution. *J. Clean. Prod.* **2008**, *16*, 1807–1813. [CrossRef]
17. Nagueira, F.; Castro, I.; Bastos, A.; Souza, G.; Carvalho, J.; Oliveira, L. Recycling of solid waste rich in organic nitrogen from leather industry: Mineral nutrition of rice plants. *J. Hazard. Mater.* **2011**, *186*, 1064–1069. [CrossRef] [PubMed]
18. Şaşmaz, S.; Karaağaç, B.; Uyanık, N. Utilization of chrome-tanned leather wastes in natural rubber and styrene-butadiene rubber blends. *J. Mater. Cycles Waste Manag.* **2019**, *21*, 166–175. [CrossRef]
19. Malek, A.; Hachemi, M.; Didier, V. New approach of depollution of solid chromium leather waste by the use of organic chelates. Economical and environmental impacts. *J. Hazard. Mater.* **2009**, *170*, 156–162. [CrossRef] [PubMed]
20. Zhou, S.; Zhou, L.X.; Wang, S.M.; Fang, D. Removal of Cr from tannery sludge by bioleaching method. *J. Environ. Sci.* **2006**, *18*, 885–890. [CrossRef]
21. Leathersmithe. Available online: www.leathersmithe.com/tanning-methods-and-the.html (accessed on 14 May 2024).
22. Dwivedi, S.P.; Dixit, A.; Bajaj, R. Development of bio-composite material by utilizing chrome containing leather waste with Al2O3 ceramic particles. *Mater. Res. Express* **2019**, *6*, 105105. [CrossRef]
23. Cardona, N.; Velásquez, S.; Giraldo, D. Characterization of Leather Wastes from Chrome Tanning and its Effect as Filler on the Rheometric Properties of Natural Rubber Compounds. *J. Polym. Environ.* **2017**, *25*, 1190–1197. [CrossRef]
24. Tahiri, S.; DeLaGuardia, M. Treatment and valorization of leather industry solid: A review. *J. Am. Leather Chem. Assoc.* **2009**, *104*, 52–67.
25. The European Commission. EUR-LEX. The European Commission. 18 December 2014. Available online: https://eur-lex.europa.eu/legal-content/EN/TXT/?uri=CELEX:32014D0955 (accessed on 6 November 2023).
26. Stefan, D.S.; Bosomoiu, M.; Constantinescu, R.R.; Ignat, M. Composite polymers frim leather waste to produce smart fertilizers. *Polymers* **2021**, *13*, 4351. [CrossRef] [PubMed]
27. Chen, L.; Tan, K.; Wang, X.; Chen, Y. A rapid soil Chromium pollution detection method based on hyperspectral remote sensing data. *Int. J. Appl. Earth Obs. Geoinf.* **2024**, *128*, 103759. [CrossRef]
28. Velusamy, M.; Chakali, B.; Ganesan, S.; Tinwala, F. Investigation on pyrolysis and incineration of chrome-tanned solid waste from tanneries for effective treatment and disposal: An experimental study. *Environ. Sci. Pollut. Res.* **2019**, *27*, 29778–29790. [CrossRef]
29. Achmad, R.T.; Budiwan, B.; Auerkari, E.I. Effects of Chromium on Human Body. *Annu. Res. Rev. Biology.* **2017**, *13*, 1–8. [CrossRef]

30. Amata, A.I. Chromium in Livestock Nutrition: A Review. *Glob. Adv. Res. J. Agric. Sci.* **2013**, *2*, 289–306.
31. Shanker, A. Chromium toxicity in plants. *Environ. Int.* **2005**, *31*, 739–753. [CrossRef] [PubMed]
32. Fendorf, S. Surface reactions of chromium in soils and waters. *Geoderma* **1995**, *67*, 55–71. [CrossRef]
33. Thatoi, H.N.; Dal, B.; Das, N.N.; Pandey, B.D. Chemical and microbial remediation of hexavalent chromium from contaminated soil and mining/metallurgical solid waste: A review. *J. Hazard. Mater.* **2013**, *250–251*, 272–291.
34. Ocak, B. Film-forming ability of collagen hydrolysate extracted from leather solid wastes with chitosan. *Environ. Sci. Pollut. Res.* **2018**, *25*, 4643–4655. [CrossRef] [PubMed]
35. Chaudhary, R.; Pati, A.; Subramani, S. A review on management of chrome-tanned leather shavings: A holistic paradigm to combat the environmental issues. *Environ. Sci. Pollut. Res.* **2014**, *21*, 11266–11282.
36. Agency for Toxic Substances and Disease Registry. 18 October 2023. Available online: www.atsdr.cdc.gov/spl/index.html#2022sp (accessed on 18 October 2023).
37. Chrysochoou, M.; Johnston, C. Reduction of Chromium(VI) in Saturated Zone Sediments by Calcium Polysulfide and Nanoscale Zerovalent Iron Derived From Green Tea Extract. In *GeoCongress 2012: State of the Art and Practice in Geotechnical Engineering*; ASCE 2012; Geotechnical Special Publication: Oakland, CA, USA, 2012; p. 3959.
38. Zayed, A.M.; Terry, N. Chromium in the Environment: Fators Affecting Biological Remediation. *Plant Soil* **2003**, *249*, 139–156. [CrossRef]
39. Verma, S.K.; Sharma, P.C. Current trends in solid tannery waste management. *Crit. Rev. Biotechnol.* **2022**, *43*, 805–822. [CrossRef]
40. de Matos, E.; Scopel, B.; Dettmer, A. Citronella essential oil microencapsulation by complex coacervation with leather waste gelatin and sodium alginate. *J. Environ. Chem. Eng.* **2018**, *6*, 1989–1994. [CrossRef]
41. Sharf, A.; Gasmeleed, A.; Musa, A.E. Extraction of Chromium Six from Chrome Shavings. *J. For. Prod. Ind.* **2013**, *2*, 21–26.
42. Yang, J.; Shan, Z.; Zhang, Y.; Chen, L. Stabilization and cyclic utilization of chrome leather shavings. *Environ. Sci. Pollut. Res.* **2018**, *26*, 4680–4689. [CrossRef] [PubMed]
43. da Costa Cunha, G.; Peixoto, J.A.; de Souza, D.R.; Cruz Romao, L.P.; Macedo, A.S. Recycling of chromium wastes from tanning industry to produce ceramic nanopigments. *Green Chem.* **2016**, *18*, 5342–5356. [CrossRef]
44. He, B.; Du, Y.; Xu, H.; Ma, J. Synthesis of Ceramic Pigments with Chromium Content from Leather Waste. *Trans. Indian Ceram. Soc.* **2021**, *2021*, 103–109. [CrossRef]
45. Majee, S.; Halder, G.; Mandal, T. Formulating Nitrogen-Phosphorous-Potassium enriched organic manure from solid waste: A novel approach of waste valorization. *Process Saf. Environ. Prot.* **2019**, *132*, 160–168. [CrossRef]
46. Wang, X.; Yue, O.; Liu, X.; Hou, M.; Zheng, M. A novel bio-inspired multi-functional collagen aggregate based flexible sensor with multi-layer and internal 3D network structure. *Chem. Eng. J.* **2020**, *392*, 123672. [CrossRef]
47. Yang, M.; Li, Y.; Dang, X. An eco-friendly wood adhesive based on waterborne polyurethane grafted with gelatin derived from chromium shavings waste. *Environ. Res.* **2022**, *206*, 112266. [CrossRef] [PubMed]
48. Younas, H.; Nazir, A.; Bareen, F. Application of microbe-impregnated tannery solid waste biochar in soil enhances growth performance of sunflower. *Environ. Sci. Pollut. Res.* **2022**, *29*, 57669–57687. [CrossRef] [PubMed]
49. Yilmaz, O.; Kantarli, I.C.; Yuksel, M.; Saglam, M. Conversion of leather wastes to useful products. *Resour. Conserv. Recycl.* **2007**, *49*, 436–448. [CrossRef]
50. Zhou, C.; Zhou, H.; Yuan, Y.; Peng, J.; Yao, G.; Zhou, P.; Lai, B. Coupling adsorption and in-situ Fenton-like oxidation by waste leather-derived materials in continuous flow mode towards sustainable removal of trace antibiotics. *Chem. Eng. J.* **2021**, *420*, 130370. [CrossRef]
51. Senthil, R.; Kavukcu, S.B.; Cakir, S.; Turkmen, H. Utilization of various solid leather wastes for the production of blended bricks. *Clean Technol. Environ. Policy* **2022**, *24*, 1889–1901. [CrossRef]
52. Zainescu, G.; Deselnicu, V.; Constantinescu, R.; Georgescu, D. Biocomposites from tanned leather fibres with applications in constructions. *Leather Footwear J.* **2018**, *18*, 203–206. [CrossRef]
53. Jeyasingh, J.; Philip, L. Bioremediation of chromium contaminated soil: Optimization of operating parameters under laboratory conditions. *J. Hazard. Mater* **2005**, *118*, 113–120. [CrossRef]
54. Sethunathan, N.; Meghraj, M.; Smith, L.; Kamalideen, S.P.B.; Avudainayagam, S.; Naidu, R. Microbial role in the failure of natural attenuation of chromium (VI) in long-term tannery waste contaminated soil. *Agric. Ecosyst. Environ.* **2005**, *105*, 657–661. [CrossRef]
55. Cavington, A. *Tanning Chemistry: The Science of Leather*; Royal Society of Chemistry: Cambridge, UK, 2009.
56. Pati, A.; Chaudhary, R. Soybean plant growth study conducted using purified protein hydrolysate-based fertilizer made from chrome-tanned leather ysate-based fertilizer made from chrome-tanned leather. *Environ. Sci. Pollut. Res.* **2015**, *22*, 20316–20321. [CrossRef]
57. Taylor, M. Effect of processing variables on ash content of galable and hydrolyzed portein products isolated from treatment of chromium leather waste. *J. Am. Leather Chem. Assoc.* **1993**, *88*, 358–367.
58. Scopel, B.; Baldasso, C.; Mettmer, A.; Santana, R. Hydrolysis of Chromium Tanned Leather Waste: Turning Waste into Valuable Materials—A Review. *J. Am. Leather Chem. Assoc.* **2018**, *113*, 122–129.
59. Masilamani, D.; Madhan, B.; Shanmugam, G.; Sarvanan, P. Extraction of collagen from raw trimming wastes of tannery: A waste to wealth approach. *J. Clean. Prod.* **2016**, *113*, 338–344. [CrossRef]
60. Bhagwat, P.K.; Dandge, P.B. Collagen and collagenolytic proteases: A review. *Biocatal. Agric. Biotechnol.* **2018**, *15*, 43–55. [CrossRef]

61. Said, N.S.; Sarbon, N.M. Physical and Mechanical Characteristics of Gelatin-Based Films as a Potential Food Packaging Material: A Review. *Membranes* **2022**, *12*, 442. [CrossRef] [PubMed]
62. Yilmaz, H.; Sanlier, S.H. Preparation of magnetic gelatin nanoparticles and investigating the possible use as chemotherapeutic agent. *Artif. Cells Nanomed. Biotechnol.* **2013**, *41*, 69–77. [CrossRef] [PubMed]
63. Wang, L.; Li, J.; Jin, Y.; Chen, M. Study on the removal of chromium(III) from leather waste by a two-step method. *J. Ind. Eng. Chem.* **2019**, *79*, 172–180. [CrossRef]
64. Ferreira, M.; Almeida, M.; Pinho, S.; Santos, I. Finished leather waste chromium acid extraction and anaerobic biodegradation of the products. *Waste Manag.* **2010**, *30*, 1091–1100. [CrossRef] [PubMed]
65. Bizzi, C.; Zanatta, R.; Santos, D.; Giacobe, K.; Dallago, R.; Mello, P.; Flores, E. Ultrassound assisted extraction of chromium from residual tannd leaher: An inovative strategy for the reuse of waste in tanning industry. *Ultrason. Sonochem.* **2020**, *64*, 104682. [CrossRef]
66. Tian, Z.; Wang, Y.; Wang, H.; Zhang, K. Regeneration of native collagen from hazardous waste: Chrome tanned leather shavings by acid method. *Environ. Sci. Pollut. Res.* **2020**, *27*, 31300–31310. [CrossRef]
67. Brown, D.A. Investigation of carboxylic acids for the extraction of chromium (III) from the leather waste and the posiible re-use of the extracted chromium tanning industry. *Environ. Technol. Lett.* **1986**, *7*, 289–298. [CrossRef]
68. Wang, L.; Chen, M.; Li, J.; Jin, Y.; Zhang, Y.; Wang, Y. A novel substitution—Based method for effective leaching of chromium (III) from chromium- tanned leather waste: The termodynamics, kinetics and mechanism studies. *Waste Manag.* **2020**, *103*, 276–284. [CrossRef] [PubMed]
69. Popiolski, A.; Dallago, R.; Steffens, J.; Mignoni, M.; Venquiaruto, L.; Santos, D.; Duarte, F. Ultrasound-Assisted Extraction of Cr from Residual Tannery Leather: Feasibility of Ethylenediaminetetraacetic Acid as the Extraction Solution. *ACS Omega* **2018**, *3*, 16074–16080. [CrossRef] [PubMed]
70. Pedrotti, M.; Santos, D.; Cauduro, V.; Bizii, C.; Flores, E. Ultrasound-assisted extraction of chromium from tanned leather shavings: A promising continous flow technology for the treatment of solid waste. *Ultrason. Sonochem.* **2022**, *89*, 106124. [CrossRef]
71. Wionczyk, B.; Apostoluk, W.; Charewicz, W.; Adamski, Z. Recovery of chromium (III) from waste of uncolored chromium leaters. Part I. Kinetic studies on alkaline hydrolytic decomposition of the waste. *Sep. Purif. Technol.* **2011**, *81*, 223–236. [CrossRef]
72. Hinojosa, J.A.B.; Marrufo, L. Optimization of Alkaline Hydrolysis of Chrome Shavings to Recover Collagen Hydrolysate and Chromium Hydroxide. *Leather Footwear J.* **2020**, *20*, 15–28. [CrossRef]
73. Pahlawan, I.F.; Sutyasmi, S.; Griyanitasari, G. Hydrolysis of leather shavings waste for protein binder. In Proceedings of the International Conference on Green Agro-industry and Bioeconomy, Malang, Indonesia, 18–20 September 2019.
74. Ferreira, M.J.; Almeida, M.F.; Pinho, S.C.; Gomes, J.R.; Rodrigues, J.L. Alkaline hydrolysis of chromium tanned leather scrap fibers and anaerobic biodegration of the products. *Waste Biomass Valoriz.* **2014**, *5*, 551–562. [CrossRef]
75. Poulopoulou, V.; Katakis, D.; Vrachnou, E. A method of removal chromium from tanned leather wastes. *J. Air Waste Manag. Assoc.* **1998**, *48*, 846–852. [CrossRef] [PubMed]
76. Ding, W.; Xuepin, L.; Zhang, W.; Shi, B. Dechroming of chromium containing leather waste with low hydrolysis degree of colagen. *J. Soc. Leather Technol. Chem.* **2015**, *99*, 129–133.
77. Asava, A.; Sang, P.; Onyuka, A. Recovery of Collagen Hydrolysate from Chrome Leather Shaving Tannery Waste through Two-Step Hydrolysis using Magnesium Oxide and Bating Enzyme. *Soc. Leather Technol. Chem. J.* **2019**, *103*, 80–85.
78. Dettmer, A.; Oliveira dos Santos, R.M. Protein extraction from chromium tanned leather waste by Bacillus subtilis enzymes. *J. AQEIC* **2014**, *65*, 93–100.
79. Qiang, X.H.; Feng, H. Collagen Extracted from Chrome Shavings Using Alkali and Enzyme. In Proceedings of the 2011 International Conference on Remote Sensing, Environment and Transportation Engineering, Nanjing, China, 24–26 June 2011.
80. Tahiri, S.; Mohamed, B.; Albizane, A.; Massaoudi, A. Extraction of proteins from chrome shavings with sodium hydroxide and reuse of chromium in the tanning process. *J. Am. Leather Chem. Assoc.* **2004**, *99*, 16–25.
81. Pecha, J.; Barinova, M.; Kolomaznik, K.; Nguyen, T.N. Technological-economic optimization of enzymatic hydrolysis used for the processing of chrome-tanned leather waste. *Process Saf. Environ. Prot.* **2021**, *152*, 220–229. [CrossRef]
82. Rahaman, A.; Raihan, M.; Islam, M.D. Recovery of chromium from chrome shaving dust. *Eur. Acad. Res.* **2017**, *4*, 9441–9448.
83. Pollet, B.; Ashokkumar, M. *Introduction to Ultrasound, Sonochemistry and Sonoelectrochemistry*; Springer: Berlin/Heidelberg, Germany, 2019.
84. Mason, T.J.; Cobley, A.J.; Graves, J.E.; Morgan, D. New evidence for the inverse dependence of mechanical and chemicaleffects on the frequency of ultrasound. *Ultrason. Sonochem.* **2011**, *18*, 226–230. [CrossRef]
85. Tadeo, J.L.; Sanchez-Brunete, C.; Albero, B.; Garcia-Valcarcel, A. Application of ultrasound-assisted extraction to the determination of contaminants in food and soil samples. *J. Chromatogr. A* **2010**, *1217*, 2415–2440. [CrossRef]

Contents

seawater to uniformly degrade the entire sample containing a UVA (UV-326). EDM has proven effective in homogeneously degrading polymers such as polypropylene in a short duration [24]. The sulfate ion radical ($SO_4\bullet^-$) was chosen as the initiator due to its conversion to hypochlorite (ClOH) upon reaction with chlorine ion in seawater. ClOH, with an extended lifetime, penetrates deep into the polymer matrix before dissociating into radicals and initiating degradation (autoxidation). This approach ensures simultaneous degradation progression from the polymer's interior and surface, synergistically accelerating MP formation. $SO_4\bullet^-$ generates OH• and ClOH, overcoming inhibitory effects and expediting the degradation process. Figure 1 illustrates the schematic representation of simultaneous and uniform degradation progression in the polymer matrix using the EDM. The behavior of UVA in each polymer during degradation was intricately examined using the EDM.

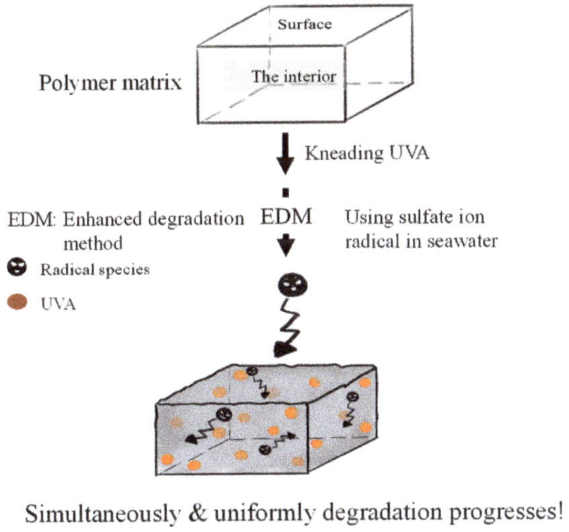

Figure 1. Schematic scheme of simultaneous and uniform degradation progression in polymer matrix using EDM.

Figures 2 and 3 depict the color changes in PP and PS samples containing 5-phr UVA film using the EDM, respectively. Samples degraded with the EDM exhibited more pronounced whitening than those degraded with a degradation initiator combining sulphate radicals and pure water. SEM images after 15 days of EDM degradation revealed numerous grooves or crushed particles. In pure water, radical species cannot diffuse into the interior of the polymer, restricting degradation to the surface, similar to weathering degradation. In addition, minimal surface changes indicate a considerably slower degradation rate compared to EDM. Notably, the difference between PP and PS lies in the whitening rate during EDM degradation. Chemi-crystallization, occurring when a polymer chain of semi-crystalline polymers such as PP is broken due to degradation [25], causes volume shrinkage and the observed grooves in SEM photographs of PP samples. It is well-known that semi-crystalline polymers such as PP and HDPE undergo morphological changes, including increased crystallinity and whitening, as the degradation progresses [25,26]. The higher whitening rate in the PP sample is attributed to chemi-crystallization, while the amorphous nature of PS does not undergo such chemical crystallization, resulting in a slower whitening rate.

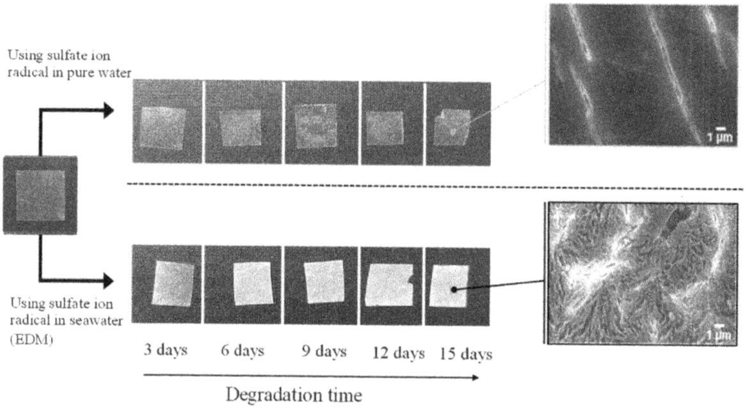

Figure 2. Color change of PP containing 5-phr UVA film using sulfate ion radicals in pure water and in seawater (EDM).

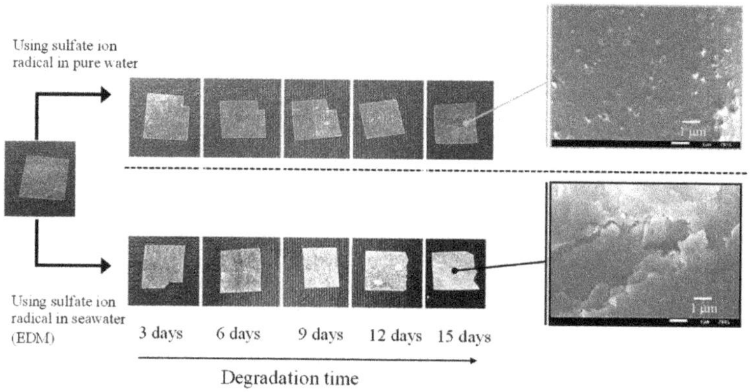

Figure 3. Color change of PS containing 5-phr UVA film using sulfate ion radicals in pure water and in seawater (EDM).

3.2. UVA Bleaching Behavior

Figure S3 shows the degradation time dependency of residual UVA in various polymer samples using sulfate ion radicals in pure water. The order of residual UVA amounts in each polymer, with increasing degradation time, was as follows: PS > PLA > HDPE> PP > LDPE. This order appeared to correlate with the glass transition temperature (Tg) except for PP and HDPE [27]. The UVA loss rate for each polymer was calculated by differentiating the residual amount concerning degradation time, as summarized in Figure S4. PS, with the highest Tg among these polymers, exhibited a UVA loss rate close to zero, independent of degradation time. Other polymers displayed similar behaviors in the change in the UVA loss rate concerning degradation time. The rate reached a maximum within the initial 3 days of degradation and sharply dropped to nearly zero after 6 days, with some variation. The scattering degree seemed linked to crystallinity levels because it was much lower for PS and LDPE with no or low crystallinity. The UVA loss behavior during progressive degradation would be closely associated with mobility of polymer chain. Higher polymer chain mobility, resulting in a lower Tg as the polymer chain easily transitions from a glassy state, indicates more substantial degradation. The presence of a crystalline part hinders the propagation of degradation and the movement of polymer chains. Therefore, if the influence of the crystalline part can be suppressed, the correlation between the UVA loss rate and Tg—in other words, the mobility of the polymer chain—becomes clearer.

Figures 4 and 5 show the degradation time dependence of UVA residual amount and loss rates in various polymer samples using the EDM. Although polymer degradation is commonly limited to the amorphous part, the EDM exhibits the capability to degrade both crystalline and amorphous components [24]. As shown in Figure S5b,d, molecular weight curves of PP and PS samples shifted towards the lower molecular weight with increasing degradation time while maintaining a consistent shape. The reduction behavior of molecular weight was similarly observed in samples of other polymers. The uniformity in molecular weight reduction suggests that EDM-induced degradation progressed uniformly with minimal distinction between crystalline and amorphous parts. In fact, for the crystalline polymers, such as PP, HDPE, LDPE, and PLA, the variation in UVA residuals at different degradation times significantly improved (see Figure 4). Figure 5 further demonstrates the degradation time dependence of UVA loss rate in these crystalline polymers, showing remarkable similarity. Figure S6 provides a comparison of degradation time dependence of UVA residual in PP samples, revealing differences between degradation in pure water, sulfate ion radical in pure water, and EDM. The equilibrium of UVA leaching from the polymer matrix can be explained by a physicochemical model in the absence of degradation. The leaching behavior of several UV stabilizers, including UV-326, from various polymer matrices has already been studied by Do et al. [28]. According to their report, the leaching behavior of UVA from LDPE in an acrylic nitrile/water mixture showed a positive deviation from Raoult's law and negative deviation for PS. Since PP is an aliphatic hydrocarbon polymer as well as LDPE, its UVA deviation value can be regarded as nearly equal to that of LDPE. The positive deviation means that there is no intermolecular interaction between PP and UVA, while the negative value for PS indicates the presence of one [28]. The significant increase in the initial loss rate observed during the EDM-induced degradation in crystalline polymers such as PP, HDPE, LDPE, and PLA is attributed to homogeneous and rapid fragmentation of the polymer matrix, leading to an increased surface area due to microplasticization [24]. A similar increase in microplasticization rate occurred in PS, but the intermolecular interaction between UVA and PS resulted in a less pronounced increase in loss rate compared to other polymers.

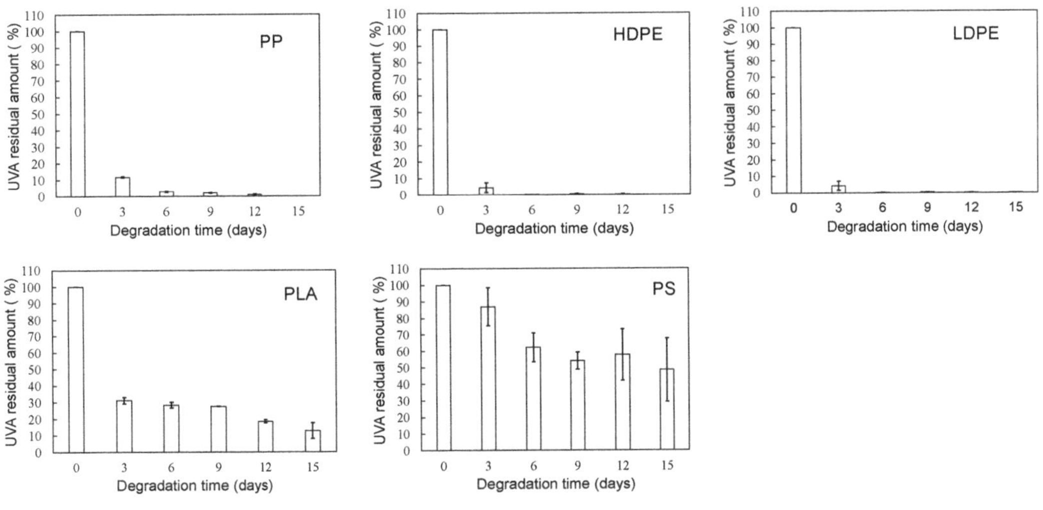

Figure 4. Degradation time dependence of UVA residual amount in various polymer samples using sulfate ion radical in seawater (EDM).

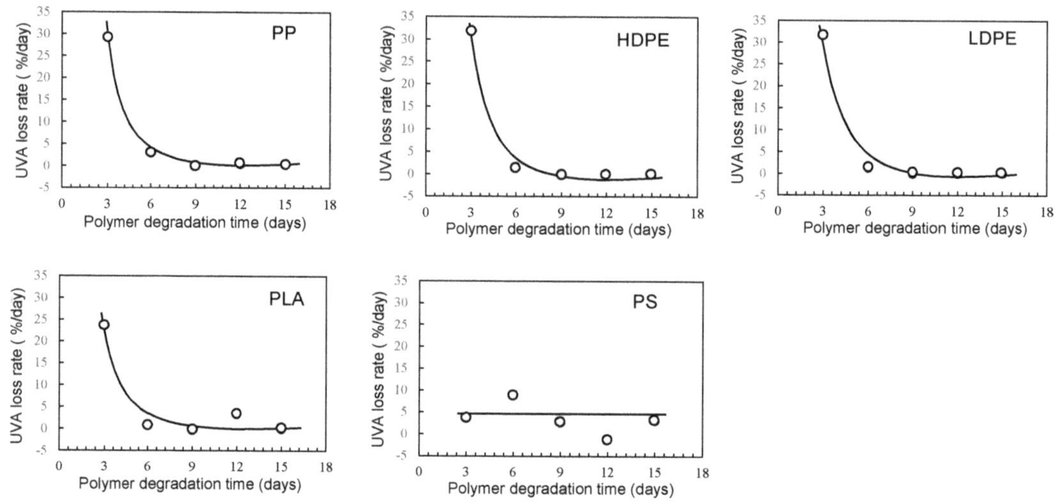

Figure 5. Degradation time dependence of UVA loss rates in various polymer samples using sulfate ion radical in seawater (EDM).

Figure 6 illustrates the scheme of UVA (UV-326) destabilization through autoxidation. The hindered phenol segment of UVA-326 acts as an antioxidant [29,30]. Polymers with aliphatic main chains, such as PP and PS, undergo degradation via autoxidation [31,32], leading to the generation of peroxide radicals (ROO•) from the aliphatic polymer main chains. As shown in Figure 6, the phenol part in the UVA reacted with the ROO•, and hydrogen transfer transpired from the phenol to the peroxide radical [29]. The resulting transformation product, quinone, was formed by reaction with hydroperoxide via the intermediate phenoxyl radical, as shown in Figure 6. The quinone part is an unstable compound and contributes to further alteration, ultimately leading to the decomposition of UVA. This decomposition process requires contact with the peroxide radical, a factor influenced by the polymer chain mobility. UV-326, possessing an aromatic ring, interacted with the benzene ring of PS through π-π stacking. If some of the aromatic rings are converted to unstable quinone rings due to UVA autoxidation, as shown in Figure 6, the interaction would decrease, thereby increasing the UVA leaching rate. Figure 7 shows the FT-IR spectra of 0- and 15-day degraded PP samples using the EDM are similar to those of the hydroperoxide group. In the IR spectrum of the 15-day degraded PP sample, a hydroperoxide-derived peak is observed around 3400 cm^{-1}, confirming that PP is degraded by the EDM. In addition, the CI value of 15-day degraded PP increased to approximately 0.6, which also confirms that the degradation reaction (autoxidation) progressed with increasing degradation time, as shown in Figure 8. Figures S7 and S8 show the IR spectra changes of PP containing 5-phr UVA and PS containing 10-phr UVA films, respectively, degraded using the EDM. The quinone group in benzophenones exhibited a vibrational mode in the spectral region of 1650–1680 cm^{-1} in FT-IR measurements [22,33,34]. After 15 days of degradation of PP containing 5-phr of UVA with the EDM (see Figure S7), peaks assigned to quinone groups were not observed in the relevant spectral region. This result supports the assumption mentioned above that the loss of UVA additive was not due to quinone deterioration but to increased solutes from outside the system, due to the increased surface area resulting from matrix degradation-induced fragmentation. HDPE, LDPE, and PLA, which are crystalline and do not interact with the UVA, are similar to PP in that the UVA loss during EDM-induced degradation is not due to deterioration but to an increased leaching rate associated with fragmentation. Even when degradation using the EDM was performed by increasing the UVA dosage in PS, which exhibited a slow UVA leaching rate, no quinone group peaks appeared until 24 days of degradation, as shown in Figure S8.

These results indicate that UVA (UV-326) deterioration in the polymers does not or rarely occurs with degradation, and the primary cause for the loss is leaching out of the system.

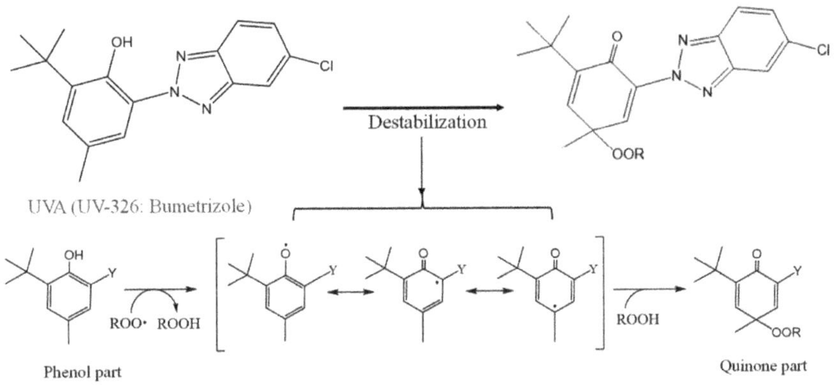

Figure 6. Schematical of UVA (UV-326) destabilization by autoxidation.

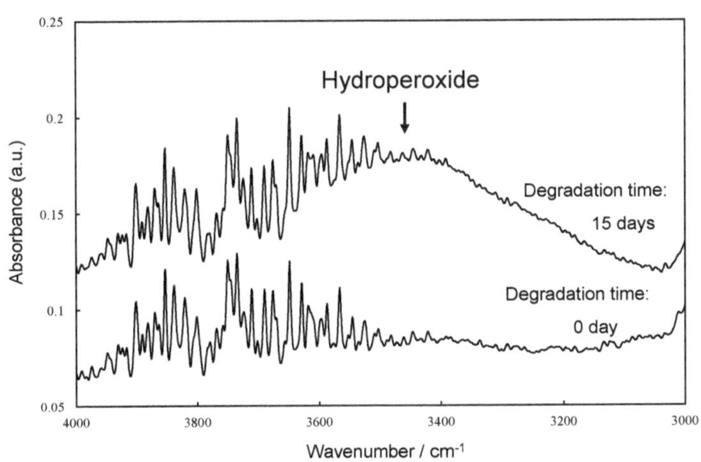

Figure 7. FTIR spectra of 0 and 15 days degraded PP samples by EDM at around hydroperoxide group.

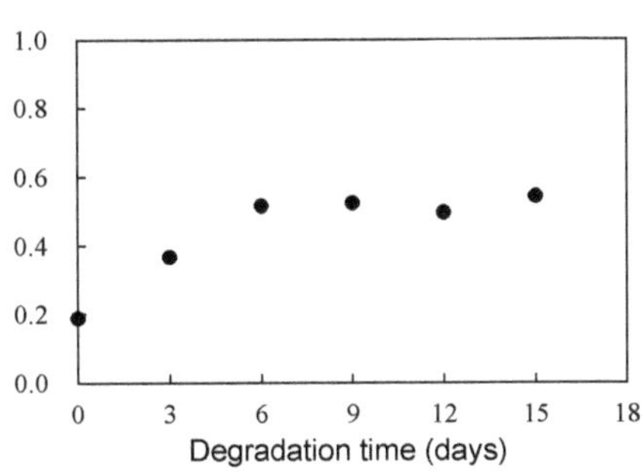

Figure 8. Degradation time dependence of carbonyl index (CI) in PP sample degraded by EDM.

4. Conclusions

To elucidate the evolving behavior of UVA over time during degradation, we employed EDM to homogeneously degrade the entire polymer sample containing it. The PP and PS samples containing 5-phr UVA films underwent rapid whitening, and their SEM photographs revealed numerous grooves or crushed particles after 15 days of EDM degradation. Notably, there was a difference in the whitening rate between PP and PS during EDM degradation. Chemi-crystallization occurred in semi-crystalline polymers such as PP, causing volume shrinkage and the generation of grooves during degradation. The higher whitening rate observed in the PP sample was attributed to the degradation caused by chemi-crystallization. The UVA loss rate in PS, with the higher Tg, was considerably slower compared to the crystalline polymers (PP, HDPE, LDPE, and PLA). Except for PS, the crystalline polymers exhibited similar behavior in the change in UVA loss rate with respect to degradation time. The significant increase in the initial loss rate observed during the EDM degradation of crystalline polymers was attributed to homogeneous and faster fragmentation of the polymer matrix, leading to increased surface area due to microplastics formation. A similar increase in microplasticization rate was observed in PS, although the intermolecular interaction between UVA and PS did not result in as pronounced an increase in loss rate as observed in other polymers. The UVA (UV-326) has one aromatic ring, which interacted with the benzene ring of PS through π-π stacking. The interaction caused a decrease in the UVA leaching rate. The phenolic part of the UVA was able to react with the degraded polymer, likely resulting in the formation of a quinone compound. As a result, the UVA may undergo degradation, possibly leading to a decrease in residual amounts, a factor other than leaching. However, no quinone group peaks were observed in the FT-IR measurements of EDM degradation of PP and PS containing UVA. These results indicate that the UVA remained unaltered or underwent minimal changes within the polymers. This study concludes that leaching from the system was the primary cause of the loss.

Supplementary Materials: The following supporting information can be downloaded at: https://www.mdpi.com/article/10.3390/polym16020293/s1, Figure S1: Pyrolysis gas chromatography (Py-GC/MS) charts of PP samples containing various UVA concentration; Figure S2: Relationship between UVA concentration in various polymers and area ratio obtained from pyrolysis gas chromatography (Py-GC/MS) measurements: Number of measurements = 3 times; Figure S3: Degradation time dependence of UVA residual amount in various polymer samples using sulfate ion radical in pure water; Figure S4: Degradation time dependence of UVA loss rates in various polymer samples using sulfate ion radical in pure water; Figure S5: Molecular weight curve changes of PP samples during degradation time: (a) by sulfate ion radical in pure water. (b) by sulfate ion radical in seawater (EDM). (c) by sulfate ion radical in pure water. (d) by sulfate ion radical in seawater (EDM); Figure S6: Comparisons of degradation time dependence of UVA residual amounts in PP samples in pure water, using sulfate ion radicals in pure water, and using sulfate ion radicals in seawater (EDM): All degradation temperatures = 65 °C; Figure S7: IR spectra changes of PP containing 5-phr UVA film through EDM degradation; Figure S8: IR spectra changes of PS containing 10-phr UVA film through EDM degradation.

Author Contributions: Methodology, H.N.; Formal analysis, H.N., T.U., S.M., A.T.N.D., H.-J.K., M.Y. and Y.K. Investigation, H.N. and T.U., Writing—original draft, H.N.; Writing—review and editing, preparation and editing, A.T.N.D., H.-J.K., M.Y. and Y.K.; Supervision, H.N. All authors have read and agreed to the published version of the manuscript.

Funding: This work was funded by the Environment Research and Technology Development Fund No. 1MF-2204 of the Environmental Restoration and Conservation Agency, provided by the Ministry of the Environment of Japan; by the Grant-in-Aid for Scientific Research No. 23K04822 from the Japan Society for the Promotion of Science; by Taihei Environmental Science Center Co., Ltd.; and by the financial support of the Nagasaki University organization for marine science and technology and for function enhancement program of research.

Institutional Review Board Statement: Not applicable.

Informed Consent Statement: Not applicable.

Data Availability Statement: Data are contained within the article.

Conflicts of Interest: The authors declare no conflicts of interest.

References

1. Barnes, D.K.A.; Galgani, F.; Thompson, R.C.; Barlaz, M. Accumulation and Fragmentation of Plastic Debris in Global Environments. *Phil. Trans. R. Soc. B* **2009**, *364*, 1985–1998. [CrossRef] [PubMed]
2. Andrady, A.L. Microplastics in the Marine Environment. *Mar. Pollut. Bull.* **2011**, *62*, 1596–1605. [CrossRef] [PubMed]
3. Jambeck, J.R.; Geyer, R.; Wilcox, C.; Siegler, T.R.; Perryman, M.; Andrady, A.; Narayan, R.; Law, K.L. Plastic Waste Inputs from Land into the Ocean. *Science* **2015**, *347*, 768–771. [CrossRef] [PubMed]
4. Halle, A.T.; Ladirat, L.; Gendre, X.; Goudouneche, D.; Pusineri, C.; Routaboul, C.; Tenailleau, C.; Duployer, B.; Perez, E. Understanding the Fragmentation Pattern of Marine Plastic Debris. *Environ. Sci. Technol.* **2016**, *50*, 5668–5675. [CrossRef]
5. Rummel, C.D.; Jahnke, A.; Gorokhova, E.; Kűhnel, D.; Schmitt-Jansen, M. Impacts of Biofilm Formation on the Fate and Potential Effects of Microplastic in the Aquatic Environment. *Environ. Sci. Technol. Lett.* **2017**, *4*, 258–267. [CrossRef]
6. Law, K.L. Plastics in the marine environment. Annu. *Rev. Mar. Sci.* **2017**, *9*, 205–229. [CrossRef]
7. Michels, J.; Stippkugel, A.; Lenz, M.; Wirtz, K.; Engel, A. Rapid aggregation of biofilm-covered microplastics with marine biogenic particles. *Proc. R. Soc. B* **2018**, *285*, 203–1211. [CrossRef]
8. Isobe, A.; Iwasaki, S.; Uchida, K.; Tokai, T. Abundance of non-conservative microplastics in the upper ocean from 1957 to 2066. *Nat. Commun.* **2019**, *10*, 417. [CrossRef]
9. Tanaka, K.; Takada, H.; Yamashita, R.; Mizukawa, K.; Fukuwaka, M.; Watanuki, Y. Accumulation of plastic-derived chemicals in tissues of seabirds ingesting marine plastics. *Mar. Pollut. Bull.* **2013**, *69*, 219–222. [CrossRef]
10. Guadagno, L.; Naddeo, C.; Vittoria, V.; Camino, G.; Cagnani, C. Chemical and morphologial modifications of irradiated linear low density polyethylene (LLDPE). *Polym. Degrad. Stab.* **2001**, *72*, 175–186. [CrossRef]
11. Gensler, R.; Plummer, C.J.G.; Kausch, H.H.; Kramer, E.; Pauquet, J.R.; Zweifel, H. Thermo-oxidative Degradation of Isotactic Polypropylene at High Temperatures: Phenolic Antioxidants versus HAS. *Polym. Degrad. Stab.* **2000**, *67*, 195–208. [CrossRef]
12. Taguchi, Y.; Ishida, Y.; Tsuge, S.; Ohtani, H.; Kimura, K.; Yoshikawa, T.; Matsubara, H. Structural Change of a Polymeric Hindered Amine Light Stabilizer in Polypropylene during UV-irradiation Studied by Reactive Thermal Desorption-gas Chromatography. *Polym. Degrad. Stab.* **2004**, *83*, 221–227. [CrossRef]
13. Fauser, P.; Vorkamp, K.; Strand, J. Residual Additives in Marine Microplastics and Their Risk Assessment—A critical review. *Mar. Pollut. Bull.* **2022**, *177*, 113467. [CrossRef] [PubMed]
14. Tarafdar, A.; Lim, J.Y.; Kwon, J.H. UV Stabilizers can FosterEarly Development of Biofilms on Freshwater Microplastics. *Environ. Pollut.* **2022**, *315*, 120444. [CrossRef] [PubMed]
15. Tian, Z.; Zhao, H.; Peter, K.T.; Gonzalez, M.; Wetzel, J.; Wu, C.; Hu, X.; Prat, J.; Mudrock, E.; Hettinger, R.; et al. A Ubiquitous Tire Rubber–derived Chemical Induces Acute Mortality in Coho Salmon. *Science* **2021**, *371*, 185–189. [CrossRef] [PubMed]
16. Lattimer, R.P.; Hooser, E.R.; Layer, R.W.; Rhee, C.K. Mechanisms of ozonation of N-(1,3-dimethylbutyl)-N′-phenyl-p-phenylenediamine. *Rubber Chem. Technol.* **1983**, *56*, 431–439. [CrossRef]
17. Nakatani, H.; Muraoka, T.; Ohshima, Y.; Motokucho, S. Difference in Polypropylene Fragmentation Mechanism between marine and Terrestrial Regions. *SN Appl. Sci.* **2021**, *3*, 773. [CrossRef]
18. Lambert, S.; Wagner, M. Formation of Microscopic Particles during The Degradation of Different Polymers. *Chemosphere* **2016**, *161*, 510–517. [CrossRef]
19. Julienne, F.; Delorme, N.; Lagarde, F. From Macroplastics to Microplastics: Role of Water in The Fragmentation of Polyethylene. *Chemosphere* **2019**, *236*, 124409. [CrossRef]
20. Julienne, F.; Lagarde, F.; Delorme, N. Influence of The Crystalline Structure on The Fragmentation of Weathered Polyolefines. *Polym. Degrad. Stab.* **2019**, *170*, 109012. [CrossRef]
21. Nakatani, H.; Kyan, T.; Muraoka, T. An Effect of Water Presence on Surface Exfoliation of Polypropylene Film Initiated by Photodegradation. *J. Polym. Environ.* **2020**, *28*, 2219–2226. [CrossRef]
22. Nakatani, H.; Ohshima, Y.; Uchiyama, T.; Motokucho, S. Degradation and Fragmentation Behavior of Polypropylene and Polystyrene in Water. *Sci. Rep.* **2022**, *12*, 18501. [CrossRef] [PubMed]
23. Mikdam, A.; Colina, X.; Minard, G.; Billon, N.; Maurin, R. A kinetic model for predicting the oxidative degradation of additive free polyethylene in bleach desinfected water. *Polym. Degrad. Stab.* **2017**, *146*, 76–94. [CrossRef]
24. Nakatani, H.; Ohshima, Y.; Uchiyama, T.; Motokucho, S.; Dao, A.T.N.; Kim, H.J.; Yagi, M.; Kyozuka, Y. Rapid Oxidative Fragmentation of Polypropylene with pH Control in Seawater for Preparation of Realistic Reference Microplastics. *Sci. Rep.* **2023**, *13*, 4247. [CrossRef] [PubMed]
25. Craig, I.H.; White, J.R.; Kin, P.C. Crystallization and Chemi-crystallization of Recycled Photo-degraded Polypropylene. *Polymer* **2005**, *46*, 505–512. [CrossRef]
26. Rabello, M.S.; White, J.R. Crystallization and melting behaviour of photodegraded polypropylene–1. Chemi-crystallization. *Polymer* **1997**, *38*, 6379–6387. [CrossRef]
27. Brandrup, J.; Immergut, E.H.; Grulke, E.A.; Abe, A.; Bloch, D.R. *Polymer Handbook*, 4th ed.; Wiley: New York, NY, USA, 1999; Volume 89.
28. Do, A.T.N.; Ha, Y.; Kang, H.J.; Kim, J.M.; Kwon, J.H. Equilibrium leaching of selected ultraviolet stabilizers from plastic products. *J. Hazard. Mater.* **2022**, *427*, 128144.
29. Klemchuk, P.P.; Horng, P.L. Transformation Products of Hindered Phenolic Antioxidants and Colour Development in Polyolefins. *Polym. Degrad. Stab.* **1991**, *1–3*, 333–346. [CrossRef]

30. Turkovic, V.; Engmann, S.; Tsierkezos, N.; Hoppe, H.; Ritter, U.; Gobsch, G. Long-Term Stabilization of Organic Solar Cells Using Hindered Phenols as Additives. *ACS Appl. Mater. Interfaces* **2014**, *6*, 18525–18537. [CrossRef]
31. Billingham, N.C. Localization of Oxidation in Polypropylene. *Makromol. Chem. Macromol. Symp.* **1989**, *28*, 145–163. [CrossRef]
32. Audouin, L.; Gueguen, V.; Tcharkhtchi, A.; Verdu, J. "Close loop" Mechanistic Schemes for Hydrocarbon Polymer Oxidation. *J. Polym. Sci.* **1995**, *A 33*, 921–927. [CrossRef]
33. Cyran, J.D.; Nite, J.M.; Krummel, A.T. Characterizing Anharmonic Vibrational Modes of Quinones with Two-Dimensional Infrared Spectroscopy. *J. Phys. Chem. B* **2015**, *119*, 8917–8925. [CrossRef] [PubMed]
34. Hellwig, P. Infrared Spectroscopic Markers of Quinones in Proteins from the Respiratory Chain. *Biochim. Biophys. Acta Bioenerg.* **2015**, *1847*, 126–133. [CrossRef] [PubMed]

MDPI AG
Grosspeteranlage 5
4052 Basel
Switzerland
Tel.: +41 61 683 77 34

Polymers Editorial Office
E-mail: polymers@mdpi.com
www.mdpi.com/journal/polymers

www.ingramcontent.com/pod-product-compliance
Lightning Source LLC
Chambersburg PA
CBHW062042220926
51850CB00010B/2641
* 9 7 8 3 7 2 5 8 4 7 0 1 3 *